SONGS *of the* GORILLA NATION

SONGS *of the* GORILLA NATION

My Journey Through Autism

DAWN PRINCE-HUGHES, Ph.D.

SOUVENIR PRESS

First published in USA by Harmony Books, New York, New York
Member of the Crown Publishing Group
a division of Random House Inc.

This British edition first published 2004 by
Souvenir Press Ltd
43 Great Russell Street, London WC1B 3PD UK
by arrangement with Random House Inc.

Design by Barbara Sturman

ISBN 0 285 63713 4

Printed and bound in Great Britain by
Creative Print and Design Group (Wales), Ebbw Vale

Acknowledgments

First, I thank Susan Arellano of Yale University Press, who asked me to write this book and took an abiding interest in it from the beginning. Second, I thank my agent, Jenny Bent, who treated this project like a child, became my dear friend, and made the book so much more than it was when we started. Thanks also to Shaye Areheart, my editor, for all her help and kindness.

I also thank my family: my mother, Joyce, my father, Ron, my sister, Davina, and my partner, Tara. My extended family—Gregg, Holly, Darcy, Mom and Dad Hughes, Hud, Linda, and Aris—all have played parts in the writing.

I appreciate my friends who gave me advice and support for the completion of this book: John Jordy, Suzanne Paola, Joyce and Susan, Stephanie, Kathy, Rhys, and the Lovejoys. And my mentors: Jane Goodall, Billy Karesh, Ben Abe, and Marilyn Smith.

ACKNOWLEDGMENTS

I thank the creators of *Star Trek* and Jeri Ryan for giving life to the character of Seven of Nine: where the gorillas showed me how to begin to embrace my humanity, Seven showed me where to draw the line and constantly reminded me that I'm beautiful the way I am.

I thank the Woodland Park Zoo, Western Washington University, and Random House for their professional support.

Mostly, I thank the gorillas.

Contents

How Can I Keep from Singing?

My life goes on in endless song
above earth's lamentations,
I hear the real, though far-off hymn
that hails a new creation.

Through all the tumult and the strife
I hear its music ringing,
it sounds an echo in my soul.
How can I keep from singing?

When tyrants tremble in their fear
and hear their death knell ringing,
when friends rejoice both far and near
how can I keep from singing?

In prison cell and dungeon vile
our thoughts to them are winging,
when friends by shame are undefiled
how can I keep from singing?

TRADITIONAL SHAKER HYMN

SONGS *of the* GORILLA NATION

Introduction

This is a book about autism. Specifically, it is about my autism, which is both like and unlike other people's autism. But just as much, it is a story about how I emerged from the darkness of it into the beauty of it. It is about how I moved full circle from being a wild thing out of context as a child, to being a wild thing in context with a family of gorillas, who taught me how to be civilized. They taught me the beauty of being wild and gentle together and as one.

What does it mean to be autistic? There are two types of autism, which I will detail in Chapter 2, but my form is called Asperger's Syndrome, and it is characterized by difficulties in processing stimuli, sensory oversensitivities, and challenges in social interaction. Though I now know what I am and believe I know what caused me to be that way—I have an official diagnosis of autism and have used that information to find coping strategies that give me and those around me a measure of peace—I will always be autistic and it will always manifest itself. I am both proud and discouraged when people say "You're autistic? Wow! I would never have known." I am glad that I am so successful at appearing normal (whatever that is), but I also wish at times people knew how hard I work at it. So much goes on that other people can't see.

I count numbers in my head or curl my toes over and over while I am talking to someone. When I am not drawn in by another person's choice of topic, I often start thinking of things that I am more interested in and don't hear anything they say. I get a physical thrill when I encounter symmetry: I love the lines and color of tennis courts and love to run on them; I love driving through tunnels and being surrounded by their roundness. When I get homesick and cry, it is because I miss times and places and not necessarily individual people. I continue to have "sense addictions": I smell all the purple irises I can when I go for walks; I still love to smell tin boxes of Band-Aids. I love the feel of having my scalp massaged and my arms tickled lightly more than traditional forms of physical contact. In times of stress I revert to eating the same thing at the same time of day for weeks at a stretch. I wear dark glasses and earplugs for the same reason. I startle and must fight rage when someone touches me unexpectedly, and I still have a very hard time with groups of people. My social awkwardness, though controlled, will always make interaction difficult for me.

Yet I believe autism can be a beautiful way of seeing the world. I believe that within autism there is not only the group—the label—but the individual as well; there is strength in it, and there is terror in its power. When I speak of emerging from the darkness of autism, I do not mean that I offer a success story neatly wrapped and finished with a "cure." I and the others who are autistic do not want to be cured. What I mean when I say "emergence" is that my soul was lifted from the context of my earlier autism and became autistic in another context, one filled with wonder and discovery and full of the feelings that so poetically inform each human life. When I emerged, I had learned—from the gorillas—far better how I could achieve these things.

I went forward by going backward. I went backward in time into the most primal and ancient part of myself. Back into the quiet recesses of the mind, where evolution has paused to breathe, bringing its people with it. I did this with the first and best friends I ever had: a family of captive gorillas, people of an ancient nation. These gorillas, so sensitive and so trapped, were mirrors for my soul as it struggled behind bars, gawked at by the distorted faces of my world, taken out of a context that was meaningful and embracing. They taught me songs about themselves, about meaning and context, about the world, and about me.

Because gorillas are subtle and unthreatening, I was able to look at them, to watch them, in ways I had never been able to do with human people. Through this process I learned that persons are more than chaotic knots of random actions; I learned that they have feelings, needs for one another, and valuable perspectives, and that as people we are reflected in *one another.* Because the gorillas were so like me in so many ways, I was able to see myself in them, and in turn I saw them—and eventually myself—in other human people.

Gorillas, like autistic people, are misunderstood. They are seen as ugly, as caricatures of fully formed humanity, as unfinished or trapped in an anachronistic world that has no value. Prejudices about what it means to be a person necessarily exclude those who are not bright on the stage of common action; those who do not welcome the glare of shining, blinding smiles, who do not lean closer to hear the roar and macramé of shouted words, who do not cut themselves and mold their flesh and spirit to fit the narrow human path, funneling upward without looking back. Autistic people can be left behind, hunted and haunted, looking through an often opaque glass.

I remember hearing once when I was young that glass is actually a liquid, moving imperceptibly like some ancient sea that separates here from there. You can look through it and not realize it's moving, but all the time the view one has through it is slowly changing. I now know that this isn't actually true, that glass is solid, but I think autistic people and gorillas know some deeper truth about this. I am always aware of a moving sort of glass between me and the world, my present and my heritage, what is seen and what is not seen and only felt. My glass is not like the wall that other people seem to have, a wall that divides and gives no view. The gorillas know this too about their own glass. They see what is beyond it. There are human people there, that is true, but the glass is also a window to their past and their future. I watch them, these slow-moving gorilla people, and they seem to intuit that if they move slowly enough, the glass will stop and show its hidden holes. Perhaps they believe they will find some kind of passage that others don't see; perhaps they feel that this way they can go home.

I found a way to go home through the glass—the glass of

my reality as an autistic person, certainly, but even more I found a way through the glass of a common zoo exhibit. The first time I knew that the glass was moving was a day, like the many days I sat with the gorillas, when we began to know one another. I knew the glass was moving when a gorilla touched me. A gorilla touched me, and I connected to a living person as I had never done before.

It was a busy morning at the zoo, where I had been able to get a job after a long period of despair, depression, and homelessness. Working at the zoo had become my lifeline, and recently I had been allowed to start working more closely with the gorillas as part of a doctoral program I was starting. They had just had their annual medical checkups, and several samples collected that morning had to be cataloged and delivered to the zoo's animal health department. Because the keeper was in a hurry and much needed to be done, she asked me to feed some strawberries to Congo, a huge silverback weighing over five hundred pounds who had recently joined us at the zoo. She showed me very carefully how to lay the strawberries on the edge of the windowsill, between the bars, and how to keep my hands back so that he did not grab my finger.

When she left to deliver the samples, I looked through the bars at the massive gorilla sitting in his corner. I could smell his body, tart and pungently sweet, drawing me in and wrapping around my own body. As I shook the tin bowl full of strawberries, he rose up like a great dark wave of spirit and flesh and hoisted himself up to sit on the ledge under the sill. He was a foot away from me. I was overwhelmed by his sheer size and presence. It was not an unpleasant feeling; it was like lying in the silent dark in the arms of a mighty and compassionate god.

He grunted and nodded his head at the bowl, raising his eyebrows to ask me to share the berries with him. I began placing the berries between the bars, careful to stay ahead of him as he quickly flicked them into his hand and popped them into his wide mouth.

Congo was quite fond of the berries and ate them as fast as I could place them between the bars. Intent on my task and compelled to put the berries in the same repeating order between each of the bars, I didn't realize he was catching up to me and eating the berries faster than I was putting them down. And then, in an instant, it happened. We put our fingers down at the same time. His gigantic finger, black and leathery, soft and warm, rested on my own digit. We stared at our fingers, and neither of us moved. Finally, I looked up into his soft brown eyes. They were dancing with surprise.

We stayed like that for what seemed like a long time, our fingers joining five million years of evolution and reaching out to bridge the gap of generations traveled. He leaned forward slowly until he was six inches from my face. I could feel his breath. His steady eyes peered into my soul, and he did not blink. I leaned forward and rested my forehead on the bars. Our faces were almost touching. We stared at each other, our fingers still together.

I relaxed into his touch and his nearness. *This is what it is,* I thought. *This is what it means to love and be loved. This is what it is to touch and look at another person and feel its meaning. This is what it is to not be alone in the vastness of the space we hurtle through among the coldness and the dying. This is what it is to live,* I thought.

Now people in the world are beginning to understand the glass that they call autism. As a result of the increasing awareness of autism and of the fact that it is a continuum, several books have now been written by people with autism. Some authors have been criticized because their stories do not adhere strictly to known patterns of autism. This kind of criticism often occurs when literature from direct sources of experience is too scarce; just a few books are forced by necessity to try to speak to all people's experience. Adults with autism long to see their experiences related and validated by others. I think it is a mistake to judge too quickly the existing books by autistics, for they cannot be all things to all people. This is true of my story also.

What I do want to accomplish with this story is to tell some of what other people with autism have experienced, and much of what I have experienced as a person with autism. Much like the deaf community, we autistics are building an emergent culture. We individuals, with our cultures of one, are building a culture of many.

Part One

A LIFE *Without* SONG

CHAPTER I

A Culture of One

I was standing at the gate of the zoo, looking up at the same gate I walked out of four years before. Since leaving, I had taken all that the gorillas had given me and made a life in context for myself. I had a family now, as well as friends, and a career as a researcher in a university. The gorillas had brought me full circle, having given me a way to understand the world. Now I gave my students the same opportunity to understand the world differently through the same family of gorillas that had been my own. This trip I had brought several of my students down to the Woodland Park Zoo, in Seattle, Washington, to meet the gorillas so dear to me.

I made my way through the zoo grounds, so familiar after ten years. Through the ticket booth that used to scare me, down the path that split apart, going to different worlds distilled into one. I silently said hello to the walnut trees, the blowing grasses, my slow shuffle scraping on the gritty path, kicking away layers of time and memory, shaking the dust from my feet, leaving it where it lay. Archaeology. I had dug myself up here. Now I was back, as a grown person, a whole person. I was a successful person, possessed of self, Ph.D. behind me now, preparing to introduce my new students at the university to the secret world of gorillas. I was there to look back and to look forward.

I traveled down the path, past the penguins, past the monkeys. A siamang screamed over an undercurrent of birdcalls. The air was heavy with the breath of growing things; it made a warm dark nest in my mouth and in my stomach. I passed through the thick bushes that waved to the path as it swept downward, a torrent, like a stream bed cupping thoughts of the past as they rush by. A spring thaw. I thawed here.

I let the memories wash me as their flow opened up and broadened; the path widened and offered an eddy. A pause. I was in front of the gorilla exhibit. There they were. I stopped thinking. I was home.

The gorillas regarded me. To them, I had never been away, because I had really been there once. Time is different to the gorillas. It is about being together, not about being apart. I am content to feel that kind of time, and I close my eyes and smell deeply the hot lemon smell of gorillas and the thick sweet smell of the hay. A gorilla keeper spots me and walks over, smiling. To her, I have been gone a long time.

I stand with the gorilla keeper. Though I had worked with her over the course of a decade, I am still not comfortable speaking with her, and I avert my eyes, not saying much. Perhaps I am strange to her.

The keeper spoke. "Alafia gave birth last October. The birth went very well. She's been a good mother." She paused between sentences. Perhaps she was feeling awkward with the silence. I glanced at her briefly. I was watching the gorillas and would rather not have been talking.

I was thinking about the process that went into a captive gorilla birth. Gorilla mating is a complex affair, built on rituals that sometimes last hours. The woman will begin to give the silverback man the "estrus gaze" that signals her desire to mate, but her readiness, especially if she is an inexperienced young woman, is often punctuated by discomfort with the unusually close proximity of her mate; turning her relatively smaller frame to his massive body is a process during which she needs much reassurance. She knows that if the dance suffers a misstep, the man of her desire can get frustrated and lash out. Though gorilla women are very seldom hurt physically in these exchanges, the emotional harmony of the group suffers, and the pair have to start once more from the beginning for mating to occur.

Then there is the birth. Captive births, until recently, were not frequent because gorillas need a cultural context—their families, a rich environment, the right foods, and learning traditions—in order to successfully bring children into the world. So up until the last many years births in captivity were seldom, and the birth of a baby gorilla into a nurturing environment where it could truly grow from the beginning was even more rare. The infant I watched was very lucky.

Alafia's baby, like most gorilla infants, probably weighed four or five pounds when she was born and was completely dependent on her mother. Until a baby is about six months old, it does not have the strength and coordination to start moving about away from its mother. Gorilla babies are dependent on their mothers both in captivity and in the wild for up to five years, often nursing until this late age.

This period of dependence is a pivotal one for baby gorillas. It is during this period that they learn to bond, first with their mothers, then with the others in their group, thus laying the solid foundations for a grasp of the social rules they will follow for the rest of their lives. This was the step I had missed in my own life, as my autism prevented me from making those crucial bonds from my earliest days with my family.

"I was glad I was here," the keeper offered brightly, interrupting my inner narrative momentarily. "She sat in the corner of her night room, and all at once the baby came out."

Still lost in my thoughts, I imagined the typical scene of a captive gorilla birth. In my mind's eye I saw Alafia catch the baby and, after looking completely surprised for a moment, begin to eat some of the afterbirth as she wiped the baby's tiny closed eyes and nose with her large callused hands. After some time Alafia may have gently shaken the infant. The baby would have opened her blue newborn eyes and gazed at her mother, letting out a wide yawn and stretching her tiny arms. Alafia would have put the five-pound infant to her chest, and the baby would have rooted around for a nipple, closing her eyes as she latched on.

"Good," I said to the keeper. I smiled. Several more silent moments went by.

"Have you ever experienced a birth?" she asked. Her smile told me that this was a topic to be joyful about, and I smiled back at her again.

"Just two."

"Oh?"

After some time went by I realized she might be waiting to hear about them. "My son's. And my own." I looked at her to see if she was interested.

She was laughing. "Yes, your own. I guess that goes without saying." She continued to laugh, not realizing I hadn't meant to be funny.

I thought about being born, and how I was born yet again—after a long time—with the same gorillas we stood watching. Being born was a long story.

I seem to remember my own birth and elements of the few days following. Though this may sound like a fantastic claim, it is true. My mother was skeptical when I told her this as a small child, but when I described the rooms and events in detail, she had to concede that many points of my account matched her fuzzy memories: the doctor and nurses dressed in light blue, standing in different places at different times, the doctor's black horn-rimmed glasses and the mask on his face. When I was born, I came out, and he lifted me high in the air for my mother to see me, with her red and sweaty face and faint smile; she was rocking her head back and forth; later I learned that they had drugged her. I remember being cold and the light hurting my eyes as I was carried over to a shelf attached to a cabinet of medical supplies right next to the door. My feet were dipped in ink and pressed to a piece of paper. I remember being wheeled to the nursery and the fluorescent lights in the hallway fading in and out in an undulating rhythm as they passed above me while I was pushed along. I remember my mother's bed near the corner far from the door in her recovery room.

When I close my eyes, I can play it back like a three-dimensional tape, replete with the smells, the sensations, and my feelings about it. I have always had this photographic or eidetic memory, and all of my many recollections of the past have a quality that makes them seem almost more real than the present. They allow me to tell the story of my life.

My Earliest Memories

Some things that I remember seem small to me: I had scarlet fever before I was three months old, I had an unusually strong reaction to bright lights and loud noises, I had no oral stage (the thought of putting objects in my mouth revolted me), and I did not like to be held. When people tried to cuddle me I would stiffen and push away from them, feeling like I was drowning. It was worse when they tried to kiss me, and their faces loomed above me and blocked out the sun.

There were a few exceptions. If I initiated this kind of closeness, I could bear it for a while, and I often approached my favorite uncle in this way. I loved his smell, like hair pomade and hot skin. He was a fat man, and I loved the security of his big arms. He didn't ever try to kiss me that I remember, and he often looked away from me when I was in his arms, which I think made him less threatening.

At about ten months I started to walk, without going through a crawling stage. I didn't like the feel of the carpet or floor under my hands, and I stood holding on to furniture to avoid it. Somehow I knew that to get around without touching the floor with my hands, I would have to use my feet, and so I began tottering between the furniture until I could walk without holding on to anything. Though I was adept at learning this skill, I would continue to be "awkward but determined," in my mother's words, for the rest of my life.

Before the age of one I was talking quite a bit, and it was not long before I could combine my new skills and walk to my uncle and say, "Ankle David, can I see your siiiilllverrr doooollllaaarrr?" I had become fascinated with the silver dollar that he wore on a chain around his neck and would ask to be picked up to see it many times a day. I loved the way it spun on its chain and threw light off its perfectly polished surface. I loved it too, because for me it was a living part of him. It didn't ever occur to me that it could be taken off as a separate part of his body.

The way I said "siiiilllverrr doooollllaaarrr" was only the start of a profound pattern I had throughout childhood of saying words and phrases in peculiar, experimental ways and having a complete fascination for words. When I was young, my parents and I often stayed at my mother's parents' house, and my favorite uncle lived there also. A hallway ran around the perimeter of the house, past the dining room, the bedrooms, the bathroom, the front door, and through the living room back to the dining room, in a big circle. My favorite game (to the exclusion of all others) was to wait in the dining room for the adults to come up with a word—the more difficult the better—and then I would speed off down the hallway, in the same direction every time, either on my tricycle or on foot, repeating the word over and over.

The word would seem different somehow, taking on new properties, as I passed my cherished landmarks. "Hippopotamus!" I would say as I passed the first leg of the journey, my grandparents' bedroom, where the word would absorb the comfort of my grandparents' bed, their clothing, the beauty of my grandmother's vanity table, and the smell of cedar drawers; then on to the bathroom yelling "Hippopotamus!" where the word would absorb the smell of antiseptic, toilet bowl cleaner, baby powder, perfume, and toothpaste; through the second bedroom, where the word would absorb the light from the fixture on the ceiling, with its fascinating mobile. "Hippopotamus, hippopotamus!" I would repeat; I would veer around the corner where the front door stood across from the stairs going to the second floor, where the word would assimilate the power of the dark stairwell; I hurried on through the living room, where the television was always on, and the word would be injected with whatever scene was on the screen.

By the time I made it back to the dining room, I had learned the word and attached it to a long list of contexts. I would never forget it. I would say the word triumphantly as I rolled to a stop in the dining room once again and waited anxiously for another word to echo and learn, echo and learn, desperate to begin moving in the big circle that was the hallway, around and around.

When I had learned all the words that my family could think of, I began speeding around and around the hallway repeating conversations verbatim or singing my vast repertoire of commercial jingles. The Doublemint gum advertisements were my favorite because when they ran on TV or the radio, many elements of the commercial were repeated twice, either visually in the case of the television, or spoken on the radio. I loved the repetition and the symmetry of these commercials. I would sing them without surcease until my mother would inform me that I was driving her crazy. This often made it worse, because then I would become anxious and need to perform the ritual of singing and going in a circle even more.

Later, while observing the gorillas, I began to understand ritual and its power a bit more. I had had the advantage of watching my gorilla family in ritual activity, sometimes as a reaction to their confinement but often born of a spiritual, an aesthetic, even an educational need. At this time I learned the value and beauty of ritual. When I was a child, however, it was just an annoyance for my family.

My need for repetition extended to routes, places, and activities. When we went to the store, the cleaner, or the park, I would insist on going the same way every single time. I would silently acknowledge landmarks as the route unwound, whether they were the buildings and hills or the flowers and trees. I had memorized everything. To me, each flower, tree, building, and hill was a person, a being with its own personality and sense of agency. If I did not see it, it missed me and felt abandoned. I would panic if we did not drive or walk by it, because it would think I didn't exist anymore and would be worried. In turn, I felt like I would disappear if I were not hemmed in by the familiar and unchanging.

I would feel like I was dying—my heart would pound, my ears would ring, and my whole consciousness would go hollow—if something changed. I remember instances of buildings being torn down, trees being cut, new roads going in, and two building fires happening along my routes. It took weeks for me to recover from these things. I would cry and yell and announce my convictions regarding the basic evil of mankind. I hated the changers and the changed. To me, change was nothing less than murder.

Oftentimes I would not accept these changes, and if we passed the site of a fallen tree or a new building, I would close my eyes and remember it the way it was until we had moved on to the safety of the sacred permanent. Sometimes I would have dreams about the buildings, trees, or fields that had disappeared, and in those dreams I would hug them and tell them how much I loved and missed them.

I also wanted to keep as many of my own accoutrements as possible the same. This meant that I did not want a new toothbrush, new clothes, new shoes. I continued to drink out of my favorite blue baby bottle until I was almost four, when it was replaced with a deep violet tumbler made of aluminum. I wanted to drink only root beer at that point, and I would delight at looking into the comforting depths of the purple world inside the cup, the smell and taste of root beer like some ancient sea against my lips, blotting out the world. I was always attracted to these alternative worlds, which is why the world of the gorillas became such a powerful one for me.

When I went to the zoo for the very first time, after a long period of homelessness and despair, it became my secret world, like my early purple root beer world. The gorillas sustained me, kept me whole, and left me feeling safe and calm, as I did when I looked into that cup.

This feeling of safety and tranquillity often eluded me as a child. Though I usually wasn't scared of things that normally scared other children, I was terrified of many objects and events that did not bother them. I was petrified of store mannequins, dolls, and ventriloquist dummies. I had an irrational fear that these dolls, dummies, and mannequins were hiding in closets, in the bathroom, around corners, waiting to get me. When I went shopping for clothes with my mother, I would stare at the mannequins, waiting for them to blink or move, my heart racing and my breath ragged. Sometimes I swore that I had seen one do just that. My strategy for survival was to hide inside the clothing racks. This served several purposes: I was safe from the murderous mannequins, and the feel of the soft fabric and the clothing's dark colors (I picked my racks carefully) and the lack of light would help me to calm down. I always felt safer in the dark.

Hiding was yet another thing that later connected me to my gorilla family—when they went behind the hills to sit or seek out the little caves in the underbrush or rock to be by themselves, I would understand. Nina, the matriarch of the troop, would sometimes hide under a burlap sack, and this too reminded me of my time hiding underneath the clothes and feeling safe.

When I was young, I stayed with my grandparents on the weekends, and those were among my favorite times. I did feel safe there. At the age of four I quietly sneaked outside at night once or twice, careful to mute the wonderful sound of the slamming screen door, and made my way around the house to look in the windows. I loved to watch the people I cared about, in the house I loved, doing the normal tasks that make a life: my grandfather would be cooking at the stove, my grandmother washing dishes in the sink, my sister playing in the living room, my uncle, mother, and father talking about politics. When they didn't see me, when we were divided by walls and glass, I could let my love for them pour out freely in the safety of the dark.

I had two very strange experiences around this time. Once I saw grass clippings falling from the ceiling. Another time I was lying still in my bed soon after climbing in, and I blinked, and it was morning. It was absolutely strange, but in a normal fraction-of-a-second blink, the night had passed. I was lying in the exact same position as I was a split-second before. I wasn't groggy or disoriented, and there had been no break in my thoughts.

Now I realize I was profoundly sleep deprived, and that this alone could account for those experiences. There is no question that my sleep patterns were abnormal. Many times after lying still in my bed for hours awake, I would be surprised to see the pinkening of the sky as it greeted the chirping birds in the front yard. I would then lie and wait for the sounds of my grandparents rising, going to the bathroom, shaving, showering, and the sound of my grandmother at her vanity table getting ready for the day.

I loved these mornings. When I was about five years old, I would lie in bed—at a distance from the activity and cherishing the closeness I felt to my family—and wait for my favorite part. I knew the smells would be coming soon. My grandmother would start to sing hymns, sometimes to the music of the Christian radio station my uncle preached on. Then she would start frying bacon and eggs. I would hear my grandfather stirring biscuits; the *shoof shoof* of the spoon against bowl and batter made me feel happy, and happier still were the smell of the biscuits in the oven and the wonderful feeling of heat that permeated the house as the oven and stove top worked hard to rise to their tasks. It was a sensory smorgasbord. It was made even better by the predictability of the sequence of these things. I felt addicted to the sensations of these sights, sounds, smells, and tactile sensations (like the heat from the oven and the feel of the bedcovers).

These were only a fraction of the sensory "addictions" I had. I craved salt and would eat it straight from the shaker. I craved burnt matchheads and would suck on them wherever I could find them. I craved Alka-Seltzer for its taste and feel. I loved the smell of my grandparents' Plymouth Fury and my grandmother's purse, which billowed the pungent smells of patent leather, Anbesol, lipstick, old dollar bills, Kleenex, pencils, and perfume. I would press my old pair of moccasins to my face and inhale deeply for several minutes before putting them back into the toy box where I always kept them.

A sound like the *thrum* of a tumbler full of milkshake when it was tapped by a spoon or the Westminster chime of the clock would fill me with rapture. I also loved the theme song of the local news program; wherever I was in the house, if I heard it come on, I would run in to listen to it. The theme song would play, and the announcer would thank Gristo

Feeds as a picture of a spinning globe provided a background. Something about the convergence of these things would fill me with deep happiness.* Other sounds, though quiet, would be painful to me and make me see colors, after which I would fight a metallic taste in my mouth.

I also had a visual affinity for turquoise.† When I looked at it, I would feel little turquoise shivers run up and down my spine and hear turquoise singing in my ears. It smelled something like vanilla milkshake and tasted like the sea, with much foam and little salt. I gazed enraptured at the items in my world that were this color: my mother's trash can and portable hair dryer; the light kept on at night in my parents' bedroom (enclosed, as a bonus, in a symmetrical white ball); and those lights, also, that shone through the conical pole lamps in the living room. In my parents' candle-making kit was my treasured bag of turquoise wax grains. There were turquoise bowls. I used turquoise-handled grapefruit spoons to eat everything with, much to my mother's puzzlement. I still eat with these bowls and spoons.

At this point in my life it was the symmetry of the mechanical that I liked. Things were made to fit together in ways that always made sense, in never-failing patterns that had purpose. Machines were both reliable and aesthetic, the

*This too would continue to be a pattern throughout my life. Even as an adult I run in to watch and hear the opening of *National Geographic* specials and Rod Serling's *Night Gallery,* for example.

†Although turquoise was one of my favorite colors from the start, until I was about four, my favorite color bar none was beige. Not only was this an unusual color favorite for a toddler (I liked it because it was soothing), but I also pronounced it in a singular fashion. All of the adults in my life frequently asked my favorite color just to hear my answer. "Bhhhhhezzzzhhhhh," I would reply.

perfect blend of function and form. Looking back, I understand that I had a very developed aesthetic sense and was constantly framing the world around me with borders informed by purpose and balance.

The gorillas I came to know later also had this aesthetic sense and would often build mysterious stone cairns in their habitat—circular piles of stones that would inexplicably appear and then disappear within a day or so. I feel certain that the cairn building came from a similar kind of drive as mine—bound to ritual, but also with a deep sense of aesthetics.

Most autistic people need order and ritual and will find ways to make order where they feel chaos. So much stimulation streams in, rushing into one's body without ever being processed: the filters that other people have simply aren't there. Swimming through the din of the fractured and the unexpected, one feels as if one were drowning in an ocean without predictability, without markers, without a shore. It is like being blinded in the brightness of a keener sight. Autistic people will instinctively reach for order and symmetry: they arrange the spoons on the table, they line up matchsticks, or they rock back and forth, cutting a deluge of stimulation into smaller bits with the repetition of their bodies' movements. When I accidentally touched Congo for the first time, it was because I was so focused on lining up the berries in an ordered and symmetrical way—I didn't notice that his giant hand had caught up with mine. Paradoxically, if not for the ritualistic habits of my autism, I would never have experienced what it felt to touch and connect with another.

Given that we seek the small and manageable, it is no

surprise that many high-functioning autistic people, unable to communicate with others above the ringing swirl, shout across the canyons of reality by writing. The aesthetic wonder of cutting and tracing the lines of one's thoughts and feelings into the steady lines of permanent letters offers the tracings of keys, the thrill of high-wire words crossing so many gaps, paintings of tiny landscapes—their horizons traced out in the mountain ranges of sentences and the strata of paragraphs. There we find a peaceful world of art and order, a land we can share.

Thus, writing was my salvation. I have said in the past, and I have since heard it repeated by other autistic people, that written English is my first language and spoken English is my second. Since I was five years old, I have written all the wonderful and terrible things that I could not bear to share. It was too much to disclose in conversation, with my eyes being seared by another human being's gaze. Though it is obvious—in reading the now-yellowed pages of a life of writing with autism—that I was different from the start, I remained both strange and invisible to all those around me.

CHAPTER 2

A Chaos of Noise: Understanding Autism

I was not diagnosed with Asperger's Syndrome until I was thirty-six. For me, as for other people diagnosed well past their childhoods, the fact of my continued existence seems no minor miracle. As I look back over the painful years I spent alienated, different, disconnected, and hurting, it's hard to understand how I made it and how it took me so long to find the reason that I lived like this.

One of the many reasons it took me so long to get a diagnosis is that beliefs about what autism is and what it looks like are often very narrow and, as a result, inaccurate. This, it can be argued, is the result of media portrayals of autistic people, which come off as one dimensional and made from a single template mold. Additionally, well-meaning documentaries sometimes capitalize on this same image, either ignoring or ignorant of the great diversity among autistic people. As a result, the public at large tends to hold in its collective consciousness a certain manifestation of classic autism, Kanner's Syndrome, the salient features of which are impairments in the use of nonverbal, expressive gestures (like facial expression and body posture), an inability to form social relationships with peers, a flat affect, delayed or absent development of spoken language, impaired ability to initiate or sustain a conversation, a preoccupation with restricted patterns of interest, a compulsion to perform specific routines or rituals, flapping or twisting, and a preoccupation with parts of objects.

Since Kanner described this form of autism, however, and in spite of persistent images of autism associated with it, further evidence has illuminated the fact that autism falls along a spectrum that shades off into clinical pictures that are very difficult for people to notice in brief encounters with autistic people like me, people who, as "high-functioning" autistics, are often given a diagnosis of Asperger's Syndrome.

One might ask how an autistic person could possibly go undiagnosed until adulthood. Asperger's Syndrome only made it into the *Diagnostic and Statistical Manual of Mental Disorders*—the compendium of diagnostic criteria for all known psychological pathologies—in 1994, but there are other factors as well. As I mentioned, most high-functioning autistic people, not knowing what is "wrong" with them, develop a lifetime pattern of using their intelligence to find ways to appear normal.

Asperger's Syndrome was first recognized and documented by Hans Asperger, an Austrian psychiatrist working in the 1940s. What separates Asperger's Syndrome patients from their lower-functioning counterparts with classic autism are two criteria. First, they show no clinically significant delay in language development (using single words and communicative phrases at the appropriate developmental stages). Second, they evince no clinically significant delay in cognitive development, in learning age-appropriate self-help skills, in learning adaptive behavior (other than social interaction), or in developing curiosity about the environment.

Despite these relative advantages, Asperger's young patients still exhibited the same sets of sensory and behavioral characteristics: they lacked the ability to connect socially and to communicate effectively; they engaged in perseverative behaviors, demonstrated extremely narrow interests (to the exclusion of all other areas), and had acute sensory sensitivities and prodigious long-term memories. I certainly exhibited these behaviors: my parents were often frustrated with me because I would "walk through" or "look through" people as if they weren't there. This phenomenon had more to do with my unawareness of where my body began and ended than with awareness of other people's boundaries. It was as if I understood the edges of other people—disjointed as they sometimes were—but I myself had no such edges.

My perseverative behaviors were many. I would listen to Simon and Garfunkel records over and over again until I was made to stop. I would feel, for example, that I needed to hear a particular song seven times, and I would have a meltdown if stopped from completing this cycle. I would need to collect a certain number of lightning bugs in one evening, or the day was ruined. I would count the pulsating whir of katydids until I felt the number was right, and then I could fall asleep.

Finally, my sensory problems were also symptomatic of Asperger's. For instance, I held my hands in tight balls because I could not cope with the possibility of getting dirt on my palms. I developed a trick of picking things up using my thumb and the side of my index finger so that I wouldn't have to uncurl my hands. I could not stand the feeling of flour or dust on any part of my body, and it set my teeth on edge to hear someone wiping flour on a board or rubbing their dusty hands together. Dust between my toes was enough to send me into a full-blown rage.

Now as then, which category an autistic person falls into in terms of official diagnosis is based on the pattern of the person's speech acquisition, their general level of intelligence, and other pieces of clinically pertinent information given by the autistic person or their family and informed by early childhood symptoms. Significant but rarely discussed is the additional deciding component in the diagnosis: the discretion of the diagnostician and her or his level of familiarity with autism spectrum disorders. When I finally sought a diagnosis, it took a great deal of research to find a physician experienced enough to make the accurate assessment I needed.

Many people with Asperger's Syndrome are not only cognitively intact but are actually gifted intellectually. Many have intelligence quotients in the very superior range. Autistic people in this category often use their profound intellectual capacities and acute memory skills to learn coping strategies that help them blend in. Because high-functioning autistic people may be invisible in this way, old stereotypes are reinforced, putting these people in an impossible position: if you can learn to interact socially, go to college, hold a job, and have a relationship, you can't possibly be autistic. Not only the public but even professionals who study autism are blind to the pain and cost, the silent desperation and continued psychological struggles that high-functioning autistics undergo every single day.

Many people, again lay and professional alike, believe that all people with autism are by definition incapable of communicating, that they do not experience emotions, and that they cannot care about other people or the world around them. My experience, both personally and with others like me, is that in many cases quite the opposite is true. A significant number of autistic people who care deeply about all manner of things, and are profoundly emotional about them, share these capabilities in the privacy of their journals, diaries, and poetry. They do not show them to the world, which is too intense and often too destructive or, worse, dismissive. They do not show them to professionals, whose beliefs about the abilities of autistic people and the power they wield over their clients sometimes make them too frightening to challenge. They do not even show them to one another. And so a vast resource of knowledge about the diversity and beauty of autism rests on countless pages, like layers of archaeology, covered with the dust of fear.

Since I had the gorillas to help me, I was able to circumvent my problems and attain a Ph.D. I have a couple of friends and some treasured colleagues within my field. I have a family—a partner and a son. But even with my experience with the gorillas, I am still a person with a neurological dis/order, and like others, I have been forced to carefully cover and compensate, so that it takes other people a while to notice that I have profound difficulties—another factor that often delays diagnosis.

This strategy, so often employed by high-functioning autistic people, seems to be more successful with age. But all the autistic people I know (including myself) report that the strategy isn't perfect and never hides our uniqueness completely. Like others who seek to be what they are not, we invariably end up with secondary problems engendered by chronic anxiety. As rage and frustration are pushed below our consciousness, we suffer depression. Somatic difficulties like stomachaches and headaches and other ailments can be chronic as a result of unrelenting anxiety and the repression of coping mechanisms while trying to fit in. Painful memories of past failures to be normal, and mounting evidence of our inadequacies, our failed attempts to "fit in," dog us. Comfort comes, oddly enough, in the form of increasing compulsions and a fierce rigidity that may cover the trail leading back to their causes. By the time a high-functioning adult seeks help—and most do not—the accretion of secondary psychological problems and the exacerbation of certain autistic features are so tangled that initial misdiagnosis, like my own, seems unavoidable.

This phenomenon is made worse by our tendency, as we grow older, to try to push our painful memories aside. Our parents may do the same. This is an unfortunate reality, because accurate memories of an autistic person's childhood and the histories of our symptoms are the very key to an accurate diagnosis. Only after an accurate diagnosis of autism is made can a person begin to understand why they are the way they are and why they always have been this way; only then can they begin to heal from the past and accept the gifts they offer the future.

The restoration of spirit that I achieved through belonging—first with the gorillas, and then to a group of people like myself at long last—is no different for autistic people than it is for all other people who need companionship. It is this sense of companionship that validates one's experience from afar. It is crucial for our sense of well-being and the awakening of our potential. But it is also, after this kind of healing, essential for our emergence as individuals.

I am an individual. I am different, for reasons germane to the phenomenon of autism and reasons mundane. All that is in between and at both ends have made my life. Within these pages, an archaeology cleared of dust and fear, I talk about this life. It is the archaeology of a culture of one.

CHAPTER 3

The Silence Before Dawn

I am remembering the years in Carbondale, Illinois, when I was five, six, seven, eight. . . . The day I started kindergarten I was deeply upset. I didn't show it outright, as many of the other children did; instead I felt hollow, and the world lost its sound. My mother waved to me as she stood in front of our car, parked by the tiny little two-room building. The teacher had told us to line up to go inside for the first day, and I was confused: Where was I to stand in the line? Would people avoid touching me? Where was the order? Children screamed, cried, rushed around. They ran after their parents and begged to go home. I stood silent as the teacher tried to marshal this chaos into order. *It's all over,* I thought to myself. *My life is over.*

With a sadness beyond description I looked for something solid. I picked up a pebble at my feet. As

I ran to give the rock—so long a symbol of stability to me—to my mother, the teacher called to me sternly. It was the first of many times I was to get into trouble for doing a strange thing. Years later I came to understand better why I made this offering to my mother. In the world of the gorillas it was not a strange thing to do. One day Congo, alone with me in the backrooms of the office, tried to trade me a piece of straw for an apple. I thought it was silly. I tried to give him the apple, but he wouldn't take it until I first took the piece of straw he offered me. It seemed to me so unnecessary—until I realized he valued me and cared about my feelings. He thought it was fair that I should have something. Hay, like my pebble, was all he had. My eyes filled with tears when I finally recognized what he was doing and suddenly wholly understood that this was the fine and sacred sentiment woven in all such acts, whether they be between gorillas, human people, or one of each.

Perhaps my mother understood this too, for she took the pebble. I was thinking as loudly as I could that I loved her, that I wanted to go home, and that I knew my wish was hopeless. I thought about my mother holding the rock all day and thus holding me close to her. I wanted to be in her pocket. The teacher yelled again. I took my place in line and felt like a prisoner about to walk into confinement, not to emerge for a very long time.

Around this time I started to gain some relief by arranging things. I loved and was exceptionally good at jigsaw puzzles. I was obsessively attached to my Tinkertoys and would carefully and meditatively line them up by shape and color. I would fall apart if they were moved. I had a Playskool desk with red, green, and blue pegs that fit into the pegboard inside. I would sit at my desk and line up the red pegs in a row; then I would proceed to the green ones, saving blue, my favorite, for last. I would regard the neat rows momentarily, then start again. I had rock collections, cicada shell collections, animal bone collections. I knew exactly where everything should be, and I knew if they had been touched.

These arranging, cataloging, and gazing rituals later took on a new dimension. No longer borne of an aesthetic need for beauty and order, they reflected the fact that I was having growing anxiety problems as well. Much later I would see this kind of behavior with gorillas in captivity. They had nervous tics similar, if not identical, to mine: hair plucking, picking at scabs, scratching, rocking, chewing on themselves, and other repetitive and self-stimulating behaviors. One gorilla spun in tight, fast circles. Another bobbed her head up and down.

My own symptoms worsened with my admission into first grade. This school was larger and more imposing than the first one, farther from home and safety, and all of the problems—uneven abilities in reading, doing math, writing, and the retention of facts—I had in kindergarten were magnified. This "skill scatter," later a prominent feature of my academic career, became more pronounced. So in addition to my social isolation, I began to feel that I was drowning academically.

I retreated deeper and deeper into nature. One lasting memory I have of this time is a trip my family took to Florida. Though I remember going for glass-bottom boat rides, seeing alligators, and, even seeing the Kennedy Space Center, my favorite memory is that of taking a walk with my mother. Just the two of us. We explored a trail near our campsite and found an almost mystical glade; showers of sun rained through the green of a canopy far above our heads. The trail wound in no hurry among the feet of trees and humble, concentrating rocks. I don't remember talking. We took turns swinging on a vine that hung between two trees. We laughed.

The entire time we were there probably didn't exceed twenty minutes, but it stands out as one of my favorite memories of my mother. I think that being outside where I felt safe, the absence of dialogue, and being alone with her allowed the walls around me to disappear so that I really connected with her deeply.

When we came home from Florida, I fell into a depression. Though before this time I had vacillated between manic and trancelike states, I now became more sedentary.* I felt like one of the still, meditating stones that I had grown to admire so. I just sat. I sat behind the furnace, or in the tangle of honeysuckle in the backyard. I sat in my treehouse. I discovered a wonderful "cave" under the road beside our house, and I stayed in it for hours at a time. I sat in places where no change would occur. I hated change.

Around this time one of my cousins stepped on and broke a crab claw that I had brought back from Florida and carried everywhere. I remember hearing the sickening crunch under her foot and moving the blankets to see my crab claw in a hundred pieces, still glittering from the sand on it. Immediately after this incident I wrote my first poem:

---

*This change is clear in my family's home movies. Where before films captured me racing around or acting up (I hated the camera because it felt like an intense eye staring at me and I didn't know what to do), now an aloofness and distance show up in place of the jumping and grimacing.

well I relley loved it but
now, it is gon so next
time
I
see
it my name
*wont be Dawn**

I was thinking about dying. I hasten to say I was not contemplating suicide. I was thinking about the eventual rest of death and hoping that heaven might be a place where nothing changed.

Echoes from the Deep

I grew older. When I was about nine, I discovered the tunnel under the road and the woods surrounding it, and I finally had a place to go where I felt completely at ease. I loved to play in the stream that ran through the middle of this little copse of trees and greet the flowers that grew in bunches all around. I became obsessed with building little forts of mud and branches, and I pretended that I lived in them. I claimed this little piece of land as my own and became very protective of it. I would clean up all the litter that had been thrown from

*Later, I developed an odd pattern of capitalization in my poetry and in other writing as well. Bs, Ts, and As were nearly always capitalized, whereas Ls, Hs, and Es were indiscriminately capitalized or left lowercase. In all of the poetry included in this book, I let the pattern stand only with the first letters of the lines, as reproducing them as they were originally written was too distracting to read. I still write with this pattern.

the road. I took samples of stream water and looked at them under my father's microscope in order to determine water quality and pollutant levels. I started a "crawdad breeding program" whereby I kept eggs and hatchlings in separate buckets and, when they had grown to a certain size, released them into the least-polluted parts of the stream. I planted native species in places where garbage had kept things from growing.

A new dimension to the game was added when I began to learn from *National Geographic* magazines about early humans and the way they lived. The little woods began to take on the look of an early tribal settlement, replete with fish-drying racks, clay pottery, stick lean-tos, discarded spears, and stone tools. I would race down to my "settlement" as soon as I could get there, take off my clothes (I never liked the feel of clothes anyway), and get down to the consuming business of being prehistoric. Something about this way of being struck a deep chord within me, and I longed for simpler times when there was less noise, less color, fewer people, and less change. As I passed the summer between my second- and third-grade years immersing myself in the ancient rhythms of my internal and external worlds, I believed that this fantasy could be real for me.

My mother would tell me I was wild. Silently, I understood how true this was. I would sit and think about it sometimes and reassure myself internally. *Someday I will find my people. When I do, I will never let them go.* I knew this was true; it was a truth that calmed me as I got lost in its prophetic depths. The gorillas and I would truly come to share a ferocity of love I had never before experienced. Our similarities went beyond the need for preservation of sameness and the needs for space, hiding, drawing inward, and exploding outward. They went beyond a difficulty with the human race, sensitivities to the world around them, and stereotyping in the face of the soulessness all around. Our affinity would meet in being filled with archaic darkness and persisting memories of a time when all things were one, even in the midst of individual responsibility.

But I had a long way to go before I could feel the true joy of this connection. My happy time in the woods was not to last. I started third grade and developed severe asthma at virtually the same time.* Either of these things alone would have been difficult for me to cope with, but the combination of both, under the grinding punishment of an unsympathetic teacher, was unbearable. I believe this was among one of the worst years of my life. Not only were my previous difficulties—sensory oversensitivity, problems with math and penmanship, social ineptitude, and need for escape—still present, but the affects of asthma made everything worse because I was tired all the time. I took experimental drugs that made me vomit and my heart race. Not only did I hate the side effects of this medication, but I had a deep revulsion to the taste and color of it. When the doctors switched the delivery system to a pill form, I would hide it under my tongue or put it in my underwear or in my ear in order to sneak my way out of taking

---

*Asthma, allergies, and other immune-deficiency disorders have been proven to have a higher incidence among people with autism spectrum disorders (ASDs). Dr. Harumi Jyonouchi of the University of Minnesota in Minneapolis and colleagues analyzed the immune responses of children with ASDs, compared with a group of healthy matched controls, and found that children with ASDs produced higher levels of proinflammatory cytokines than did children without autism.

it. My parents caught on to this, so I had to keep devising more sophisticated ways of hiding the pills. When I didn't take the medication, I would be in full-blown attack, having to sit up to breathe and sleep, feeling like I was suffocating. I ate hardly anything but crackers and soda and became so exhausted from this diet and from lack of sleep that I would often be too tired to move and would just turn my head to vomit and continue to lie in it until my own exhausted mother would come clean me up.

My third-grade teacher was not moved by my plight. In general she found me arrogant, insolent, lazy, unpredictable, excitable, and loud. She thought I had been spoiled and that my problems were the fault of my parents. She would assign me more math homework than the other kids (which my father often had to pick up on his way home from work because I missed so much school) and forbade me to focus on the English and reading assignments that I loved. She would make me copy long sentences hundreds of times to improve my penmanship, which never improved as a result of this method. When I was in school, she would announce my failing math scores in front of the class.

I always did my worst on the math tests. We were to complete our times table in three minutes. She would stand beside me with the stopwatch and press down on me from above as I sat, pencil stock-still in hand, unable to move under the strain of her physical proximity and my problems with math. Once she sent me out in the hall after becoming disgusted with me for not being able to complete a multiplication table. She screamed in my face, *"Do that table!"* I couldn't move. When she came back some minutes later, I was still in exactly the same position. She put her face three inches from mine (I still remember a tiny triangular scar she had on her forehead) and screamed the same sentence again, *"Do that table!"* I was literally petrified.

When she returned yet again a long time later to find me in the same position, she screamed, *"Do that table, or I'm calling your mother!"* I broke down and sobbed, without moving. "Call my mother, please. I want my mother." She leaned back, arms folded across her chest, and informed me that she would do no such thing. My mind swam. I often couldn't take in people as whole entities, even when I was relatively relaxed. Now the threatening and disembodied pieces of my teacher swirled around me, attacking from every angle. I was caught in a whirlwind of horrible sensory information and unrelenting criticism. I needed my mother and knew that this demon, in the form of flying, taunting parts, had the power to keep her from me. I don't remember how it ended.

I never told my parents about it. It didn't occur to me that I could communicate about things that happened. I simply wasn't able to understand that use of words.

A poem I wrote at the time captures my growing sense of aloneness and the jarring reality of the world:

The Streets were hot
and The smell of smog
burned my lungs and
throat. The honking and
shouting bellowed and
hurt my ears. I said
hello to people, but

they glanced blankly
at me and went on. The buildings
Tower and look down
on me.

Another time a young woman came to the door of our classroom, and my teacher told me to go with her for some tests. I asked what kind of tests they were. In front of the whole class she answered, "We are testing you to see if you are retarded." I went with the woman to the special education room, where she asked me a series of questions about the meaning of words, about some readings she asked me to do, and some writing I had also done. She asked me to do some math problems (which I guessed at answering so that I wouldn't have to sit there unmoving) and put some puzzles together. I felt uncomfortable from the start because I had profound problems with following sequential directions, especially when they were spoken.

She asked me about my family. I told her where my parents had been born, what they looked like, and what activities they engaged in. I gave her similar statistics for my sister. I tried to bring the conversation around to topics that interested me so that I could share an intelligent moment with her, and I was irritated that she kept muscling the conversation back to these idiotic questions. I kept asking her when we could talk about something else. "Soon," she would say. This meant nothing to me as I had very poor concepts of time. It was meaningless. It was clear the examiner was not going to be any clearer regarding the matter of when I could talk about my interests. I resigned myself.

She asked me how I felt about different things, and I answered as best I could because I felt a moral obligation to do so. I remember telling her that noodles were "happy" because they went in circular motions in the pan when the water was boiling, that willow trees were "sad" because they couldn't straighten out their branches. I couldn't come up with a concept for "scared." To me this was just a general state of being.

I did take a chance and reveal to her that I felt "special" when I rode the Round-Up ride at the fair.* When she asked me for clarification, I told her, "It makes me sleepy." I didn't have the sensation-discrete vocabulary to tell her it made me feel content and relaxed as well as joyful.

She thanked me in a soft voice at the end of the session, told me I was quite an articulate young lady, and walked me back to my class. Neither I nor my parents had been informed in advance that this test was to take place, and the results were never shared with us.† The only discernible change after the test was that my teacher let me spend more time with language-based projects once again.

When we began to have social studies and I learned

---

*Going to the Du Quoin State Fair was one of my favorite experiences. I loved concentrating on all the blue things there: blue plastic ducks that you picked up to win prizes, the blue cotton candy that I wished I could live in, and the Round-Up ride topped with blue lights. The Round-Up utilized centrifugal force to press the riders into a spinning chamber. I have since wondered if the sensations of relaxation and well-being I experienced on this ride made it my version of Temple Grandin's squeeze machine—a device based on a cattle chute that applies pressure to her body in a uniform and comforting fashion.

†I have since learned, after contacting the school as an adult in order to obtain these records, that the practice of nondisclosure was common among school districts in the 1970s and before. These records, as they exist as the school's property, are often destroyed soon after the child exits the school system.

about anthropology, I was fascinated with early humans and knew that I would be an anthropologist someday. I practiced on the playground: I would make willing children lie still on the ground so that I could "discover" their remains. I would act out the stages of evolution. I would run around the playground with a notepad and ask people why they were doing things. This was something I could really understand. After all, anthropologists lived among those whose ways of being were totally foreign to them in order to learn more about their culture. Though most of my fellow children couldn't answer my questions, some gave me thoughtful answers and helped me begin to understand some facets of human behavior.

Academically, I still had what could be considered savant skills in language and reading. Words described real experiences, and their curves and lines left a mental trail for me to follow by sense memory, whereas numbers threw curves at me and stonewalled me with their lines, barring me from understanding them, where they came from, and where they went. Math did not describe anything to me; if people themselves were often disconnected parts—sometimes one, sometimes many—how could I hope to quantify the rest of the world? Discrete amounts had little meaning for me. This was complicated by the fact that I saw numbers as specific colors. To me, if you added seven, which was pine green, with three, which was dark blue, you got nine, which was a blue-green color.

When I got frustrated with any task that presented these kinds of problems, I would rush through my work, sometimes drawing all over the assignment paper and handing it in hoping to get points for originality. I never did.

A Song from Far Away

I was amazed when I started fifth grade and my teacher, Kay Eckiss, seemed to understand my problems.* She let me complete the Scholastic Reading Achievement tests at my own pace. When I finished them, she let me choose my own reading material and write book reports. She never criticized my bad penmanship. She allowed me to help start and edit a class poetry journal called *Writers on the Wing*. She allowed me to do minimal math, and if my problems resurfaced in one area, she would let me try another area altogether.

Best of all, she did not make me go outside for recess and play with the other children. We would have long talks about topics in theology, philosophy, the social sciences, and politics. She wouldn't laugh at me when I stated my belief that trusting authority and believing in God was naïve and senseless. She didn't laugh at me when I told her that I had joined Save the Whales and proceeded to school her in the horrors and unsustainability of whale hunting. She took me seriously when I said I felt like I was a million years old.

The times when she did disagree with me, she gave me *reasons*. Logical and well-thought-out ones. This separated her from almost everyone else I had ever known. She made me want to understand. I started to feel like there might be a chance for me not to be alone.

---

*Kay and I have remained friends through the years, and I am happy to say that she now teaches special education and helps children on the autistic spectrum to reach their full potential.

It was then that my parents told me that we were moving out west. They had "fallen in love" with the Rocky Mountains and wanted to leave Illinois.

One must bear in mind that in addition to finding a friend in Kay, I was still spending every weekend with my grandparents and uncle and that I was as close to them emotionally as I could be to anyone. I had a special connection with my grandparents and even felt that they were really my parents. My mother and father were so young that they seemed more like older siblings to me. This was wonderful most of the time, because they were creative and energetic. But I needed the stability of my grandparents to function. On top of that my reliance on context, my attachment to "places as people," meant that I would not derive any comfort from talking to them long-distance or through letters.

To add the final blow, we were to give up almost all of our possessions in order to travel light as we set off with no destination. My parents had no plan as to where we were going to settle down. Their idea was to try to find land in Colorado or Montana before the beginning of the next school year. For a child with autism, any one of these things—losing grandparents, moving to a new house, giving up possessions, traveling without a specific goal—can lead to a complete breakdown. I still think of this turn of events as the most traumatic in my life, and it still affects me today.

I remember very clearly going through my things and deciding what to keep and what to sell at the yard sale. Finally, I decided it would be easier to give everything away without looking at every single piece and telling it good-bye. All or nothing. At the last minute I picked out a couple of things for storage because I felt I was "betraying" certain objects by letting them go.

The house was sold. I said good-bye (literally) to the irises in the backyard, the doghouse, the garden, the underground water tank. I went down and said good-bye to my tribal settlement. I went into the cave under the road and sang a nonsensical song to express my grief. I insisted on being the last person to leave the inside of our home. I stood alone in it and said good-bye to a whole life and the living thing that was my house.

We packed up the Volkswagen camper and bought some ice for the cooler. We stopped by my grandparents' house and said good-bye to them near the garage. My uncle kept smiling strangely. My grandmother held a tissue, like a small, dead bird balled up in her hand. My grandfather held tightly to her arm, and I wasn't sure who was keeping whom from falling down, from crying, from running to catch us as we pulled out of the driveway and down the street to the wholly unknown. I was ten.

Moving Passages

After the move the only redeeming thing that stands out about this time was a herd of old Morgan horses that had been let go to roam wild in the hundred acres across the road from the trailer. My relationship with them and the emotional sustenance it gave me was a foreshadowing of the closeness I later had with the gorillas. I watched them, learned their habits, and knew where to find them at all times of day. It took me weeks, but I finally got close enough to touch one. These horses were as magical to me as if they had been the ghosts of unicorns.

I brought them apples from the orchard and vegetable scraps, and soon they let me sit among them as they grazed lazily in easy company with me and with one another. One day I stood on a stump and leaned over the top of the leader. He let me slide on, and I laid the whole length of my body along his back. I just lay there with the warmth and pressure of his massive body supporting me. I let my arms fall down along his withers. I started to do this every day. I never tried to make him go anywhere, I just wearily collapsed onto his hot, chunky muscles and let him eat and walk, eat and walk. We were both too old to want a direction. All around me I could feel the personalities of the rocks, the trees, the grass, the hills. Sometimes we would wander far from where we started, and I would have to tear myself away and walk as fast as I could to get home before dark.

Meanwhile, the energy that powered the very methods I used to escape made them impossible to maintain. They all came crashing down one day in music class. A girl who was known for talking back to teachers and bullying other children started to provoke the music teacher. The scene escalated and culminated in the teacher physically strong-arming the girl out the door. Both of them were screaming, cursing, swinging; they knocked over chairs and broke a picture on their way out. As they were headed out the door, the classroom exploded in laughter. The lights suddenly pierced me with their blinding whiteness, the cheers and laughter pounded against my ears, the field of my vision dissolved, and the last thing I remember is falling to the floor and some people trying to pick me up.

Apparently my classmates carried me to the principal's office, where the school nurse took over and put me on a cot in the back room. My father was called out from work to come and get me. When we got home, he laid me in bed and smoothed my hair. He lay down with me, and for the only time I can remember I put my head on his chest and talked to him about my fears. Nothing I said made any sense, and my normal articulateness became an incomprehensible jumble of references to real events, things I felt physically, stories I had read, and things I needed in order to survive. He was trying hard to help me but had to admit that he and my mother had considered finding a hospital for me.

Well, that's it, I thought. *I am going to find a way to be "normal" and fit in with people at school if it kills me, because another change like that certainly will.*

This was when I started drinking. I was in seventh grade. My mother's bowling friends bought alcohol for me. I remember approaching one of them, who always looked like fifty miles of bad road. She plucked an ever-present cigarette out of her mouth as she listened to me: "I am gratified to see you this evening. I trust that you are well. I am hoping you would consider buying me some wine for the following reasons." When I finished, I think she felt so sorry for the doofus before her she relented against her better judgment. She became my main supplier. Though I still had no friends at school, the other kids admired my drinking prowess and the nerve I seemed to have; I would get drunk right on the school grounds. To avoid the abuse of my classmates, I would run as fast as I could to the "drop-point" that my mother's friend and I had agreed upon as the hiding place to leave the alcohol. I would stay there, drinking, until the bell rang again. Often I would write poetry, scrawled more and more carelessly as I drank:

I heard the sound,
Of a sigh in the
Still ness of the night,
It was so soft

And so long that,
I could've written a thousand
Poems in the age of
The silence
*And in's lengh**

Te world could live
And die

And in our lives
We could not imagine
Its age

With ours.

---

*in its length

People occasionally asked me to go to parties, and I would go for the free alcohol. I would often end up being effectively raped by some boy or man who would be thwarted from carrying out a technical rape by my complete physical rigidity. I wouldn't be able to move, to speak, or even to see. In these situations people's faces would disintegrate, and I would feel like I was going to explode. I would just disappear. I would sometimes have to walk miles to get home when these men got mad at me and pushed me out of the car.

I also sometimes got into trouble with men when I mistook them for other people. This was a strange phenomenon. I always had face blindness and could not recognize people unless I knew them well or within a certain context. If I saw a man who looked like another I knew, he would take on all the characteristics of that man. One time, for example, I was at a carnival and saw a man who looked like the kind reverend at my grandparents' old church. I couldn't conceive of him having different qualities than the reverend. I chased the man at the carnival and held his arm when I caught up to him. I tried to talk to him and be physically close to him. He was just starting to touch me back and say sexually suggestive things when my parents found us and intervened. I grew increasingly uncomfortable around men, finding I was unable to trust them.

By the time I was in eighth grade I had decided to avoid everyone possible and pretend no one existed. Though I was still drinking, I was doing it alone. I had found a new passion and threw myself into it with the fervor of all that preceded it: philosophy. I had found a book detailing Kant's discourses on morality, and it set my mind on fire. I knew I must use the tool of philosophy to lead a moral life. I knew immediately that I could utilize philosophical systems to organize and make sense of my life and to solve my problems. Philosophy gave me a map and a yardstick.

My commitment to Kantian morality was put to the test when school resumed and I walked up to the doors of the high school for my first day. I was full of hope that finally my mind would be challenged and I would find a raft in a sea of intellectual apathy. I knew they must teach philosophy in high school and offer other valuable kinds of instruction. I was breathless with anticipation as I neared the doors.

A football player in his junior year stepped in front of me.

"I hear you're queer, is that true?" he bellowed in my face.

I thought back over the summer. I remembered a conversation I had with a woman I had been drinking with. I had told her that I wasn't interested in men, that I had really had it with them trying to touch me all the time, that I felt overwhelmed by them, and that I preferred the company of females and had since kindergarten. I thought about the definition of "queer" and concluded that in the broad sense I qualified under that category.

I had no overtly sexual feelings for anyone, male or female, but I had to admit a desire to be near some women because they made me feel good. This interpretation of the facts, coupled with the recent killing of Harvey Milk in San Francisco and my readings on Kant, converged and left me no choice but to answer the football player in the affirmative. I could never understand why these things mattered to anyone. I thought that with my answer we might be able to talk about this issue and I could make an important statement of my views.

He immediately hit me and knocked me down. He said, "I don't think you heard me right. Are you a *queer?*"

Failing to see why his show of brutishness would affect my answer in any way, I asserted the truth once more. He hit me again. This pattern repeated another time or two. Then the bell rang, he gave up, and we went inside. High school was downhill from there.

Notwithstanding my efforts to clarify the entire truth of matters of gender, sexuality, and experience, with their inherent complexities, I was labeled a "faggot" (sic) and had that stigma flung on top of the heap.

I wasn't impressed by my teachers, who had no love of philosophy, literature, or anything else, but merely tried to educate people for careers in logging and ranching. My grades, due to my old problems with people, emotions, and senses, coupled with the verbal and physical harassment brought on by my new status as a lesbian, continued their steady decline. Though I studied Kant, Aristotle, and Shakespeare at home, I was getting Fs in virtually all areas at school.

Because of complaints by girls who thought I would "rape them" in the locker room, I was taken out of physical education class and allowed to sit in the library until the end of the school day. This was the only place of peace I could find, for it was the only place I could really be alone.

My family moved out of the mobile home we had been living in and into a seven-by-fifteen-foot camper. We had bought twenty acres and begun the process of building our new house. My father designed the house himself on a piece of paper: supported by vertical lodgepole pine logs, the inside, except for the bathroom, would be completely open, making the interior of the house look like a forest. Between clearing the land and hauling water (we hadn't had running water in our previous trailer, and we wouldn't on the new land, either), we were tearing down a house built in the mid-1800s in order to use its wood, windows, and nails. After school I would help my family as the old sheltering mother building came apart bit by bit. It was as if she had died and her bones were now being carried away by birds and tiny animals, the rest left to the wind.

My sister and I extracted and straightened the bent, rusty nails for reuse, pounding them flat with a hammer against blocks of wood. I hated this job. What I hated more, though, was coming home to the tiny trailer with no room to move and no privacy. I would do anything to get outside. I would eat my lunch in the outhouse, I would sit in the truck and listen to music on the radio. I would walk across the land for hours. We lived in the trailer for almost two years, and even in the bitter cold that came with the Montana winters, I would go outside and not come in until I could not stand the cold any longer. I would lie in the snow and look up at the stars and wish I were floating in space where there was infinite room, nothing to touch my body, no sound, no light, no people.

The gorillas I came to know later all needed this kind of space around them, and when I began working with them, I became aware of the intricate dances they did together in order to remain intact in the group individually while keeping a bubble of private spirit around them. They made it seem effortless and unintentional. One would step forward to pick up a carrot. Casually sniffing it, the gorilla would look down its nose, poised to step lightly in response to the chain reaction that nuance would set in motion. Another gorilla would pass by, regard the carrot momentarily, then pass on toward another distant reward. Another gorilla would watch, then move in another direction to busy herself with picking at imaginary sores on her arm, glancing up and away to take in the unfolding scene. A fourth gorilla would pass through the center of these movements and circle around to start another subtle ripple of attention and inattention, his leg raised in midstride, pausing for a second, then turning his head. All the time they would stay just so close and just so far. I understood this way of being, but I could never find other people who understood its rules. It was as if human people had lost the ability to dance to this music. So my looking away seemed strange, my walking by people without seeing them, my crossing the street to go around them. To me this was as natural and old a way of being as heartbeats. In hindsight, the ways of the Gorilla Nation, so subtle and so natural to me, are in stark contrast to the ways people constantly invaded my personal space to taunt and assault me.

I remember sitting in the camper one night after dark, with the family around the table, when I told my parents about the first day of school and the series of events that led up to it. I told them about my feelings about men and women and some of the odd ways I was behaving at school. I alluded to the violence and cruelty, but I didn't tell them everything. I'm not sure why I told them what I did. In retrospect, I realize that I had a strange system regarding what I talked about and what things fell under the category of "truth duties." I didn't tell them about my drinking and other things I had done because at the time I believed that the truth would not affect them. I simply couldn't conceive of the reality of them "caring" about me and thus being worried about me.

As I told them about my feelings, my parents listened carefully and did not make much comment. Soon afterward my mother started driving me down to the nearest city of any size (an hour away) so that I could explore a group run by gay and lesbian people. I had no interest in dating for sexual reasons and still felt that sexuality ran on a continuum. In the feminist atmosphere of the late 1970s and 1980s, much was being discussed about the politics of sexuality and gender, and I was committed to searching out my own internal intimacies without outside influence. For my fourteenth birthday I asked for subscriptions to *Psychology Today* and the National Organization for Women's monthly publication.

As I read these periodicals and got to know the people in the group (who were mostly men), I grew to realize that the two-gender system presumed in our culture and aimed at purifying heterosexual sex ignored many complexities of biology and psychology, as well as the nature of changing people and their political commitments. There was room for everyone and their evolving expressions, I thought. I wondered if one could be "lesbian" in one's orientation to the world but choose to never have sex. It is important to understand that the move toward "un-labeling" gender- and sexual-preference-based identities and the "deconstruction" of roles and behaviors came much later for me; I had no access to these ideas then. If I had,

I would have realized, perhaps, that my being and my sexuality were unique and not readily classified. Obviously, having autism underpinned much of my gender identity or rather lack thereof. I have since learned that most autistic people do not see gender as an external or internal category that is important or even applicable, especially to themselves; as a result of learning about myself, I finally unlocked the enigma of sexual desire, now understanding it well. We should rid ourselves of the myth that autistic people simply don't have sexuality; rather, it is different and takes more time to unravel. I knew I did not want to choose a subservient role in a relationship, that I wanted to be an intellectual, and that I wanted someone to respect me, support me in reaching my goals. I had always associated these needs with masculinity.

I enjoyed the men at the group. Many of them were well cultured and introduced me to new realms of thought, as well as to classical music, opera, art, and literature. It was many years before I realized that these things were not the specific domain of gayness, because I learned about them in this context and I absorbed information in contextual blocks.

I loved the parties that they threw: all manner of people weaving all manner of ideas in an atmosphere free from judgment and ripe with possibility. Nothing I did or said was cause for upset. I think the other people involved cared about me and wanted to be supportive, even if at times they didn't understand me. I know they worried about my drinking, as well.

With the acceptance I was experiencing through this flamboyantly creative community and with the support of my parents,* I found it increasingly difficult to cope with the abuse at school and the rigidity—both academic and moral—that I found in the school structure itself. I stood out as a freak in school: my tics, my monologues, my sensitivities, my imperviousness to criticism and suspicion of authority, my disdain for connection and avoidance of social interaction, my political convictions, my obsessions with philosophy and anthropology, and my odd style of dressing and speaking all led to total ostracism and active aggression.

Determined to last out my high school career, I tried to find new ways to get through the day. I imagined that no one else existed in the school, so thoroughly that people literally disappeared from my sight. With my eyes fixed ahead and focused on nothing, I would wander the halls, seeing them empty, quiet, hallowed. I became unable to hear my teachers and would be speechless when they called on me.

People would corner me in the bathroom and force my head into the toilet, slam me into my locker, and throw trash at me in the hall. They hit me in the head with books and spit on me. They defaced my locker. They took my food away. Once some senior students made a sign with a derogatory word on it and hung it around my neck. I didn't take it off. I walked around with it on because they had no power over me. To me, the words that people shouted at me, their thoughts

*Though I often fought bitterly with my father, it was always clear he wanted the best for me. Like most parents of people with autism, my parents had some autistic traits of their own. I believe that it was our very different (or sometimes very similar) needs that led us to conflicts. Notwithstanding my rocky relationship with my father, he always supported my right to be a unique individual.

about me being crazy, the ways they treated me, were all as real as the sign, and I couldn't take them off. Once a thought or action has taken shape, it is real forever, and paradoxically I also think change is its own redemption. I was swimming in a sea of ugliness, hate, and intolerance—what good would it do to remove a cardboard sign? I didn't have the energy. I was drowning. This kind of behavior never failed to confuse me. The gorillas often had to face the same kind of abuse: human people would swagger up to the viewing window and stand stiffly, threateningly, staring down mercilessly at the gentle gorillas, who were too polite or too overwhelmed to meet their searing gaze. "Look at that ugly fucking gorilla!" they would say. "Jeezus! Get a load of the face on that thing!" They would draw their lips back into terrible, mocking smiles. The gorillas would sit. "Hey, John! That one looks like *you!*" and John would snarl back, or play along, or push his interlocutor. Then they would rush away to poke their twisted fingers and twisted minds at other animals or people elsewhere.

Through it all I believed in pacifism and never hit anyone, never screamed at anyone. I did grab someone once and was expelled for three days. The whole scene is surreal to me as I remember it. I didn't give any warning. I just sat, looking down at my desk, as this girl spewed forth all manner of vile epithets. When she ended with the words "fucking freak of nature," I leaped over my desk and wrapped my hands around her throat. The only difference between this event and the others is that she was whispering to me. I still don't understand why this made a difference, but it did. I was so repentant that I fasted for the duration of my expulsion. I had broken my own code, and I could not forgive myself. I do not say this to demonstrate any kind of superiority, though I do believe that trying to resolve conflict nonviolently is better than resorting to aggression. Rather, I present these years of suppressing my rage and right to physical defense as an underpinning for the lost years that followed.

CHAPTER 4

Starting Off Key

I remember clearly the day I quit school. I was sixteen. I had not been contemplating leaving. That day when I went in, I assumed it would be just one more in a series of grinding days with no end in sight. It was morning, and I was standing at my locker putting my things away. Someone came by and said something to me. It was a small remark and carried with it no imminent danger. But it was threatening.

With slow deliberation I took my things back out of my locker and walked out of the school. I called my mother on a pay phone to tell her I had left and wouldn't be going back.

Years of calmness and reason evaporated seemingly overnight. All I wanted to do was to stay with my friends in the neighboring town and drink. Suddenly I didn't care about anything at all. I had no future, no strength, and scant guiding principles. My parents were frightened for me—they could see that I had really given up. They helped me pay for a tiny room in a seedy hotel where roaches were my only (and unwelcome) companions. I shared a bathroom with the rest of the hotel, mostly men, who were always trying to corner me there. I was sixteen.

I had no life skills and no clue how to survive. I had no idea how to go about getting a job, and even if I had, I couldn't read maps well, I got lost easily, and I was terrified of talking to interviewers face to face, not to mention the fact that I had no work experience. I stayed huddled in my room, afraid to go out, afraid to move. I listened to the radio and rocked on my bed under the gray light accidentally falling through the dusty curtains that screened the view of the brick wall opposite my room:

Window pain,
Makes me sad,
It hurts,
As I gaze through the matrix,
Out on the world I used to know . . .

Separated,
I will never return,
I looked back through the window,
And the wind outside,
Pushed me away,
So cold,

It froze my tears,
I lie here,
In window pain

Sometimes I found a way to get alcohol or drugs, but I had no food. I remember someone brought me a can of kidney beans once, and I ate them off of my comb because I didn't have any utensils.

My parents tried to come to check on me as best they could and bring me things to make my life easier, but I was losing contact with them, moving away from them and into nothingness. I was ashamed that I couldn't function, and sometimes I avoided them when I knew they were coming.

Eventually the rent on my room was due, I had no resources for another month, and I was asked to leave. This was the beginning of several lost years during which I was homeless. I traveled all over the country, staying with people who took pity on me or saw something in me. At one point I met a woman who had moved back to her family's ranch out west and turned it into a gay and lesbian guest ranch. I lived there on and off for several years as the owner became a good friend. It was a great place to be in the summers, as I could live in one of the guest teepees, but it got too cold in winter, and I would have to leave. I would go anywhere with whoever I thought would keep me from living outside or starving. Some of these people were accustomed to having hangers-on: drug dealers, mothers, models, artists. I cared about many of these people, but they never really saw me, and I suspect that

I never really saw them, either. I wrote a poem for one woman I knew during this time.

she knew the way the wind
blew
when it came down the alley
she knew the way it hurt when
a friend was drunk and crying,
she knew the empty ache of a
life lost
she was old.

she read about better places once, in a
dirty paperback she found in a pile
of trash on the corner.
she could read.

somehow she knew how to hold
me in a way she had never been
held,
and she could cry
I saw her once

she cried just one tear.

Most of the people I knew during this time in my life had had enough, just like me. Many were very smart, some were brilliant. Nearly all had been born with a unique way of looking at the world and had suffered the consequences. Ours was the sad refuge of those who, abandoned by all those around them, stood with their backs together looking out on the world . . . still alone. We were too damaged, too tired, too cynical to care: about ourselves, or about one another.

I lived in a kaleidoscope those years. I was looking down a narrow tunnel at broken, colored fragments of people and dreams, turned toward a too-bright sun as I rolled from place to place, one eye blind.

In the self-centered and aloof culture of the 1980s, my social impairments and my emotional and physical distance made me appear "cool."

After touching back down at the gay vacation ranch during my travels around the country, I made friends with the ranch's cook, who was singing the praises of the West Coast; she was moving back with her partner and invited me to come along. I remember her saying I was too young to be so lost. I was really just too overwhelmed to have any direction, a situation that was the worst of many vicious cycles.

By the time I landed in Seattle, I was ripe to be taken in by the post–New Wave scene and sucked into the mass of trendy clubs replete with leather-clad dominatrixes, big-haired voguers, and tattooed modern primitives. At first glance I scarcely seemed strange among that company. In fact, though I appeared "cool," my cover was always blown as soon as someone talked to me. Then it became painfully obvious (as I was told later) that I was strange. I wanted a traditional family, I thought morality was very important, I didn't want to live immersed in excess. My thinking was very rigid. I must have been a very dissonant character indeed as I talked about the value of self-discipline and looked down at my dirty engineer boots through my sunglasses.

I was homeless again, after the kind cook and her partner grew tired of my odd habits and chronic lack of resources and asked me to leave their home. I couldn't blame them even then.

I ate out of garbage cans when friends didn't generously share their food with me, and I slept in a church stairwell at night. On a visit to my parents' house, I built a tiny diorama of their house that I carried with me everywhere. I would sit in the church stairwell and look into it, imagining I was back with my family.

I'm so far away from home,
In many more ways than one,
It's been so long since I've felt loved,
That even my longing is lonely,

And for every pair of friends I see,
And every hungry dog,
I can see my shadow linger,
Because I live in a shadow box . . .

Shadows of fear,
Shadows of love,
Of times I held up my pride,
Of times I was sick,
And cold . . .

But shadows pass to forget the pain,
Just like a camera never captures the sadness,

I'll live in my shadow box,
Like a hungry dog.

Being homeless is more difficult for autistic people than anyone I can imagine. Most homeless people I have known who are young adults had the advantage of weaving together a street family wherein they felt relatively safe and knew that others were going to help them find food, shelter, and a sense of belonging, in addition to a very real means of protection. Some of the scams I have heard about street kids using to get food takes more than one person. Sometimes they even split up their money from panhandling and share it. I couldn't even panhandle for change because it would require me to identify individual people under stress and interact with them. People certainly won't give you money if you can't even look at them. What I used to do then was to go down to the Pike Place Market and watch for people (at a distance) to throw food away. I would look for people who seemed clean and healthy. When they threw a sandwich or something in the garbage, I would go get it and take it into a bathroom, where I would go through an elaborate series of rituals: tearing away the edge of the area of food recently bitten, tapping the food three times in sets of threes, rewrapping it with the outside of the paper or napkin on the inside, then taking it to a special spot on the stairs near the entrance of the Market to eat it. I also would go to pizza places and sit in a booth with a soda until people left large portions of their pizza behind. I would hurry and get it before it was cleaned away by the person busing tables. This had to be done carefully, as the service staff would ask people to leave if they caught them doing this.

Being cold was hard. Luckily, a lot of the time I was technically homeless, caring people would let me stay at their houses in exchange for a small amount of money or in exchange for a little help around the house. When I didn't have this luxury, I slept outside. The longest period in which I did this was a several-month stretch in Seattle, when I slept in the stairwell of the church on 13th and Spring Streets. Regular street kids had other options. "Squats," or houses that are abandoned and taken over by street kids, are often fiercely defended; this takes several people pulling together to defend a territory. Even certain areas in general are defended, and I remember a few times when people yelled at me to get lost or go away. It is hard to describe the loneliness of being autistic and homeless. It is bad enough trying to deal with one's sensory overload and different reality when one has a constant safe place. I think being homeless over those times was as close to hell as I can imagine.

Sometimes I thought of all the rooms, even closets, that were dark and warm and snug in the big houses around me. I fantasized about what I would like to eat. It always seemed strange that within forty feet of where I sat lay all the peace of someone's home; I looked in windows and saw families together eating dinner, firelight and candles, laughter, even fights, and it was all an oasis, or perhaps a mirage, that I could never touch. I would wonder, *What's wrong with me? Why can't I hold a job? Why can't I have real friends and a real family?* All I was left with was the cold. I learned to sleep sitting up. It reminded me of my asthma attacks and the nights I couldn't breathe.

Siren Songs

I would talk my way into the clubs for company and warmth, but also for a more important reason: dancing. Though I had been taking drugs and drinking for years, I gave them up when I got to Seattle, and after I did so, I found the power of my body. I had never been good at sports, not only because they required intense interaction with others but because I had always been awkward. Dancing was something different, and I was good at it. For some reason, where noise and light had caused me pain before, the flashing and throbbing of the clubs pushed me deep into myself. I knew the freedom of self-expressive movement and could dance alone all night. I was always dismayed when someone asked me to dance, because to me it was an absolutely solitary activity.

I was even more dismayed when someone walked up to me once when I was dancing and put a five-dollar bill into the front of my pants. To me, dancing was not a sexual thing at all. It didn't even occur to me that people watched me. Looking back, I know people might find that hard to believe, since I always took over the highest platform on the dance floor and as a result could be seen from everywhere. I did this, however, to get away from the crowds and find a place to move freely.

When the person who had tucked the money in my pants walked away, an acquaintance of mine standing nearby told me that I really should think about dancing for money. I couldn't understand what she meant. She told me she worked in a strip club that was owned and operated by women and that the job paid very well and had great benefits. I found this idea repugnant.

"You wouldn't have to sleep out in the street anymore," she countered.

Reluctantly, I went for an audition and was hired immediately. I liked the actual room where we danced: it was about eight by fifteen feet and completely covered in mirrors, which gave it a surreal and futuristic quality. The music was good, and I could continue to dance in my own world without seeing or interacting with anyone.

Our dancing establishment, called the Amusement Center, was unique as a business. To start with, it had glass all around the stage so no one could touch the dancers. Visitors dropped quarters into a meter to get a screen to go up, for about twenty seconds a quarter. More important, though, the establishment was managed by women who were themselves dancers and knew the grueling demands of the work. They were sensitive to the physical demands of working eight-hour shifts, often in high heels, with only a ten-minute break for each hour of dancing. They tried to schedule us based on our abilities and limitations. They paid us by the hour so that we didn't have to compete with one another for money. They hired tight security guards and had enough of them to walk us to our bus or wait with us for a cab after work.

As dancers we were all allowed to choose stage names and were encouraged to explore different parts of our personalities. Since I wasn't comfortable with pretending to be someone really foreign to me, I spent a great deal of time exploring parts of myself that I had long ago abandoned. Many women stayed with traditional negligees, teddies, and fishnet stockings as their work attire, but I often came dressed in animal skins and body paint. I wove beads into my hair and tried to work animal movements and tribal dances into my dance routines. I was interested in ritual dance and dress and read a great deal about it, even on my ten-minute breaks in the dressing room. I found that I resonated with many of the insights that a dancer who considered herself pagan shared with me about the archaic roots of what we were doing. During that time I became interested in the modern primitive movement and went so far as to get several tattoos and even run for (and win) the first annual Ms. Seattle Leather Woman contest. My goal in exploring this emerging phenomenon was to learn about and honor the tribal roots of urban society; most other people involved were interested in hardcore sadomasochism, a fact that, looking back, should have been screamingly obvious to me. Though I tried to make the best of the year, it really was a debacle.

I often lost myself in my secret animal universe and forgot that customers came to see me as a dancer and not as an endangered species. I would take huge leaps across the stage, screaming in midair, and come down into a banging crouch in front of some horrified man who hadn't seen me launching toward him. Just as quickly I would leap away to grab the wooden railings situated all around the stage area and hang suspended from them, swinging around with all my might. I was called into the office many times and warned not to run amok in this way, but I just couldn't seem to help it. Thinking of it now, it probably wasn't just the customers who were complaining; the other dancers must have felt that working in a zoo exhibit was something they simply hadn't signed on for.

Maybe I should have known earlier in the whole process that I really belonged with the gorillas.

The women I worked with were all very different, and very few fit the stereotype that the public holds for erotic dancers. None of them were currently prostitutes, although some of them had been at one time. They had taken up dancing for a variety of reasons—they had children now and wanted fixed hours. They were frightened by the Green River killer who was then on a steady and punctuated mission of killing, or perhaps they were just getting older and found it too tiring to be out on the street.

Many of the women were very caring and looked out for one another, bringing doughnuts or other snacks to share. One woman, who was a devoted mother, stuffed her purse with all manner of pharmaceuticals just in case someone at work got sick. She would come into the dressing room and put her enormous bag down so that it tipped over and overflowed with Dramamine, nasal spray, headache remedies, Midol, and tampons. Flipping back her long, wavy brown mane and fussing over her latest patient, she would chastise them, saying things like "You need to stop actin' like such a damn ho—out partyin' all night. No wonder you have a damn headache." She would screw up her mouth in disgust while simultaneously getting the woman water and smoothing her brow. Her grousing was ubiquitous and comforting.

Another woman who used to fascinate me was stage-named Candy Girl. She had been working as a dancer for years before I knew her. She was small and pale but had the most beautifully defined body; like something you would see in a fitness magazine. Dancing, as I said, was not an easy job physically, and the machinations that Candy Girl excelled at ensured that every muscle of her body was hard enough to bounce a quarter off of, including each muscle surrounding her pert genitalia. One of her many amazing tricks of erotic acrobatism was to do a walkover and place her heels on the backstage mirror in a headstand, then slide her legs down into a split, whereupon she would suck air into her vagina and force it out to make a rhythmic popping sound. It was like she was blowing some kind of amazing vulvular smoke rings. I can't say I understood what it did for the men who viewed this spectacle, but it was definitely a crowd-pleaser.

I remember staring at her once when she performed one of her famous Olympic-level floor routines, which utilized each available corner of the stage, and vicariously sharing in her clear exuberance: she was happy to be alive, to have such mastery over her body, to be in control of herself and the flow of her life-generating spirit. Wanting to connect with her in some way, I called out over the blare of the jukebox, "Where did you get the name Candy Girl?" At first I didn't think she heard me, because she didn't look at me and just kept up her galloping, kicking, and jumping pace. Several moments later, still without turning to me, she said in her slow, southern drawl, "Ain't it obvious?" I suppose it was.

A much-loved and longtime worker, Lacey, dispensed gentle Christian advice to the young women around her, who were often troubled or tired. I still have an image of Lacey sitting quietly among the bustle of the dressing room and presenting such a beautiful picture; she was so serene, so accepting, and right with Christ, whom she loved more than her own breathing. She had been raised within the paradoxically freeing confines of strict morality in a black Baptist church.

One may wonder how such a religious woman had come to lead a life as a career dancer. Lacey was blessed—for so she considered it—with the most enormous breasts I had ever seen. They actually prevented her from leading a normal existence. I asked her once if she felt angry that through no fault of her own she was forced to lead what many might consider an immoral life. She seemed genuinely surprised. "The Lord give me dese," she said, as she pushed her small hands under the mountains of flesh that gave her headaches, backaches, and rashes. Lifting them up to heaven as a testament to her belief in their divine origins, she continued, "He give me dese so I could spread love. Den He give me dis job so I could get along in life." Her thankfulness was very touching to me. And spread love is what she did best. I have many memories of her pouring out her warm spirit and fulfilling her Lord's directive as she comforted those in despair, in exhaustion, or even just in a state of acute bitchiness. Laying a sweet hand on one of their naked knees, herself as naked and golden brown as a hymnal cover, she would lean forward to look into their eyes and say, "Child, just believe on de Lord and all will be well."

This kind of compassion and concern was what kept me with the business for several years. However, there were definite and awful downsides to this work. Apart from some happy exceptions—the man who used to come into one of the private booths that lined the stage and do a striptease for the dancers while grinning playfully and shaking his tiny butt, or the man who wrote poetry for whichever woman was dancing for him, sticking his little notepad full of flowery appreciation to the glass window so she could see it before the screen went down—many of the men who came to visit were indifferent or even cruel. These men would come in and start looking over the women onstage (usually four to six) in order to pick one out. If a dancer a man didn't like tried to entertain him or even inadvertently got in his way, he would shout obscenities at her and gesture wildly for her to move out of his line of vision. It used to make me angry to hear men yelling at Lacey to "get your big tits out of my way" or hear them yelling at someone's mother to "move your fat ass so I can see the pretty bitch over there." Most dancers developed a thick skin and let the words and gestures pass right by them. Exceptionally patient dancers would try to be nice to these men, believing, as they told me, that they needed the kindness of women to help them feel better and then hopefully to act better. I can't count the number of times Lacey said, "I'm sorry, sugar," to some man who was heaping abuse on her and testifying to her every perceived physical and moral flaw. It was clear to me even in my naïveté that she meant it.

Even worse than these men's rejection was their demand to inspect a dancer in whom they were interested as if she were a car they were thinking of buying. "Bend over and spread your legs," such men would say matter-of-factly. "I want to see how big your pussy is." Once a man had his pants down around his ankles and was masturbating in earnest (the inevitable end point of this dog-and-pony show), he would often bark orders at the woman: sit this way, touch yourself that way, suck on your own breasts, go kiss another dancer. He had no feeling about the woman beyond her existence as a sexual object. I reflected on this later when I read descriptions of autistic people being known for their emotional distance; but the men I saw do these things were family men who ostensibly loved their wives and children. These "respectable citizens" were doctors, lawyers, city council members. Even though I was only in the first stages of reaching beyond myself in ways that were overtly social, I tried to show that I cared for the women I worked with.

When I later spent time with the gorillas who lived behind glass, I became aware of the parallels of our existences—zoo animals and dancers on display, exposed to hateful words and twisted judgments.

Even Closer Scrutiny

During the years when I was dancing I was confronted—through the nature of the job and through the people I now came in contact with—by my expanding sexuality. Though many people probably think that women who dance are very sexual and have wild escapades with men at every given chance, many of the women I worked with who identified as straight hardly ever felt like having sex. I am guessing that after an eight-hour day of watching these men indulge in their own sexual gratification, after-hours sex was, let's say, unappealing. But I'm sure that most of them would have married any man who acquiesced to giving frequent foot massages, without asking anything in return.

On the other hand, and this isn't widely known, many

professional erotic dancers are lesbians (and were lesbians *before* getting into dancing). I would say, looking back, that anywhere from about a third to a half of the dancers working at the establishment at any given time identified as lesbian. Maybe they withstood the job better, given that they didn't have to take it home with them and have it affect their relationship dynamics in the same way it would have if they had gone home to a man. Perhaps there was a distance there that let them feel that men were barred from their intimate lives, that they therefore had less power over them in ways that seemed to get under the skin of straight women dancers. Or maybe they just had to toughen up to face the world in so many ways that dancing was the least of their challenges. I'm not sure.

For whatever reason, there were a great many lesbians working as dancers, and I worked alongside them each day. As I mentioned earlier, I had noticed during my first years on my own that my aloofness converged with the times—the "me" generation 1980s—to forge an illusion that I was very cool indeed. Another component of my "mystique" was my appearance: I wore leather jackets because their weight and thickness calmed me; dark glasses, sometimes even at night, because they cut out some of the stimulation to my nervous system; and heavy boots that made me feel secure and grounded as I clomped around in them. I must have looked like a perfect practiced stud with all the trimmings, when in reality I was withdrawn and armored primarily out of anxiety and confusion.

Again and again it happened that a woman I worked with or met out at other clubs would literally throw me up against a wall and make passionate advances to me. I was always amazed by this phenomenon, though I was proud of eventually being able to anticipate some of these attacks in advance. They certainly had a raw power of their own, and I found their intensity at turns overwhelming and gratifying. When I knew such an outburst was inevitable, I would get a kind of rush, like being touched before I actually had to contend with someone's hands on me, a direct experience that was more pleasant in the abstract than in reality.

In my limited world, though, experiencing such power made me feel like I had control over something in my life. Though I was awkward and for a long time successfully repelled the sexual advances that women made toward me, the feelings I began to have of sexual power and control made me resolve to explore the phenomenon more thoroughly. I decided to become the best object of sexual advance that any woman could dream of.

I watched erotic videos, read all kinds of manuals, listened to women talk in the dressing room offstage about the things they really liked in a lover, and grilled them about what worked and what didn't when it came to sex with women: What was their idea of perfect romance? What made them want to surrender their control so completely to another person? What was missing in each of their partners up to the present time? What did they like to hear said to them? I asked them all these questions with a desire to commit every piece of information to memory, with an intensity that surprised them.

I drew extensively on the protocols I had compiled to apply my data in successive encounters with women full of desire. My applications were successful, so much so that apparently word got around. Soon it seemed that a whole new wave of passionate women had learned that I believed:

(a)
that anywhere from four to eight hours was the acceptable length of time for engaging in sexual activity;

(b)
that absolute fulfillment of a woman's needs was my sole objective;

(c)
that it was my duty, when engaging in sex, to make my partner feel more beautiful and desirable than she had ever felt; and

(d)
that my own pleasure was irrelevant.

In point of fact, I believed that most pleasure was ultimately irrelevant; but it was clearly so important to the people around me that I knew even then that I was missing something. I conceded, from a philosophical point of view, that pleasure could make people happier and more relaxed, even ideally better thinkers and better people. I felt an obligation, from this perspective, to ensure maximum happiness on the part of my partner.

When women first began making their unexpected and

passionate advances toward me, I was still extremely naïve. Having long been a dedicated fan of literature that included themes of love and sexual passion (I read *Lady Chatterley's Lover* at twelve, for example, and it remained a favorite for many years because of its natural setting and the simplicity of true love once found), during my first sexual experiences I honestly believed that the women in question were in love with me. Part of me rejoiced in the conviction that at last I was loved for myself: somehow these women had seen the "real me" that was so invisible to the world.

Of course, this was never their intention at all. I would be proceeding under the assumption that they wanted to live together or date exclusively, but by the time I was calculating how to make our relationship progress smoothly, they had already moved on to another conquest or prop for their sexual fun. One instance particularly stands out in my mind. One of the first women I knew in Seattle—the woman who got me the audition at the dancing establishment—grabbed me by the collar while we were out at a club and started to kiss me passionately. Without protest from me, she dragged me back to her house, me dutifully following at her heels. After spending the night together, it never occurred to me that we weren't a "couple."

When I saw her dancing with someone else and kissing her in the same way a few days later, I was shocked. But I strode up to the undulating pair and asked, "So, uh, when did you think you would like me to move in?" Thinking back on the looks on their faces as their mutual fog lifted and I came into focus, they must have thought I was making a weird joke. I was, of course, completely confused and hurt. This type of gross misunderstanding happened a few times with women who mistook my habitual wearing of leather as a sign that I was interested only in sadomasochism. I was shocked and dismayed the first time a partner, in the throes of passion, asked me to harm her in some way. I was uncomfortable with these requests and struggled with what I saw as my conflicting duty to please my partner on the one hand but never add to suffering on the other. So there I would be, trying to convince them through logical argument that they should abandon their counterintuitive insistence on discomfort and "settle down" into a more conservative relationship. Offering to provide this option to them personally drew incredulous stares from these hardcore women who just wanted me to tie them up, push them to their physical limits, have sex with them, and turn off the light on my way out.

It still makes me sad to remember that most of them seemed very damaged to my soul's eye—and deeply wounded. Going forward through these experiences, I continued to mistake women's lust for love and made the innocent mistake of believing with each sexual experience that I had found someone who loved me.

This mistake was in fact predicated by several factors that continue to challenge me even now. First, I am likely to take images at face value. In most popular movies, when people seduce the object of their desire, it is because they have yearned for them until they can no longer bear to be without them. Happily ever after is a foregone conclusion. Second, though I never felt love for these women, I felt that it was somehow honorable to play it through to the end. It was as if I were so completely detached from the shared reality around me that I was moving through a dream. *Oh,* I would think to myself, *so this is what I am going to do now. I will live with this woman, and we will be a couple. Interesting.* Third, my concept of "the future" was, for all intents and purposes, completely undeveloped. If something was one way at the moment, I couldn't imagine it any other way. If a woman desperately wanted my company at a given point in time, I could never conceive of it being any other way. This same inability to see the future as developing into something different from the present was a further stumbling block in my education and other areas of growth. This lack of future concept may be responsible for some of the poor impulse control that many autistic people struggle with and that can make relationships so difficult.

The few times I had relationships that lasted more than a few weeks, this "future" problem would surface in other ways, once again snuffing out their nascent promise. If I argued with my partner, I couldn't imagine *not* fighting with her at some point in the future, and I would believe that there was no way of saving anything between us. Panicking as a result of what seemed like failure, I would have rage attacks that were, effectively, self-fulfilling prophecies.

Neurotypical humans, I have learned, have a deep fascination with, and an attraction/repulsion response to, the kind of figure I presented. I have long been an abiding fan of the Seven of Nine character in the *Star Trek Voyager* series, as I identify strongly with her. Originally a young girl who was "assimilated" by a race known as the Borg, who evolve and expand by absorbing other beings into their single-minded and efficient collective, she is molded into an organism who is part machine and part human. She thinks very rationally with the focus and harmony of the collective, which uses her talents for conciseness and intellectual efficiency to the utmost until she is "rescued" by the captain of *Voyager*. The captain is determined to "restore her humanity." The pain she subsequently endures is heartbreaking. She has no inherent understanding of social protocol, the importance of valuing the happiness of the individual over the whole, or why the isolation of individuality is prized over efficiency and pragmatism. I have related to her every falter and crushing defeat. I believe we share the same brand of innocence and wisdom.

Like Seven of Nine, I find that I am only part "human" and very much something altogether different; I am overwhelmed by the social demands of "normal life"; and I am lonely. I often don't understand why people do the things they do, like the things they like, and remain unaware of the grossest social contradictions. I don't understand the strange rituals associated with attraction or why people are obsessed with certain physical characteristics.

The objectification of Seven always upsets me, by those who pay heed only to the tightness of her unitard biosuit and the way it accentuates her trim, statuesque, and culturally predictable form. Everyone seems to be rooting for her to become some kind of mindless sex-toy. Her cool directness, her obvious intellectual brilliance, and her sensitivity generate in them only a desire to see her lose these traits in a flood of uncontrolled lust. Instead it should inspire them to see the beauty of her retained distance.

For the same reason, looking back at my own early experiences upsets me. People seemed to want to conquer the very parts of me that made me initially appealing to them. For people to fantasize about toppling and ravaging the very qualities that make a person such as me or Seven truly beautiful seems irrational. Like Seven's character must have felt, I have many times seen no end in sight to the eternal drifting through cold space in a ship out of control, without the comfort of a living mirror for my soul or the warmth of a companion who loves me because she understands me from the inside of my body out.

Much of this took place before the gorillas entered my life. My heart was still not open. To me sex was an exercise to be mastered. A thing open to quantitative assessment, improvement—perfection. My goal in these situations was to perform flawlessly, to be efficient, to afford my partner no opportunity for frustration or complaint. A couple of times early on in this great experiment it crossed my mind that if I had just fulfilled a woman's every desire, perhaps I should have an opportunity to fulfill my own—which, rather than sexual, was intellectual in nature. In one instance a woman was lying in my arms after hours of sex, and we were watching the sun come up, and I said, "Do you know what Plato says about forms?" She answered, "No. But I know what to say about your form." Pushing her away to block a kiss, I said, "No, really. What do you think about the idea that there are timeless blueprints for things?" She twisted her mouth in disgust. "You're weird," she said.

So I continued to engage in these activities without sharing anything, without feeling anything, even though at times my boredom, mixed with upswellings of panic at such close physical contact, made me feel like some kind of mechanical device. Maybe that's all they wanted. Most of the time I didn't care.

But sometimes I longed for someone to find the key that would finally bring me out of the sexual machinery that I had built around my emotions and make me feel. I wanted someone to help me forget where I was; to help me let go. I wanted to be in love with one special someone and give them my heart and my life between breaths and between heartbeats.

I wanted to close my eyes and lose myself, to give in and give myself over to someone who loved me, really loved me, for me. I wanted to see God. I wanted to see God in the face of my lover, in my own body for a few brief moments, to see God throbbing, pulsing, and living everywhere, the way other people did. All they would have needed to do was to touch my hand gently and mean it sincerely.

But it never happened. God was never revealed.

I kept dancing.

Subtle Strains

Another good thing about the dancing job: the management was very strongly in favor of education. The managers attended rallies for the rights of workers in the "sex industry" and had started college funds for some of the women there (and gave them health and dental coverage). I greatly admired one woman in particular who was dancing her way through medical school. I didn't know what I wanted to do with my life, but I admired the women who were learning.

Right about then something wonderful happened, something I never would have expected. I hated the city and longed for nature. Since I was small, it was the only place I was truly comfortable. When I was homeless and broke, I had tried to talk my way into the local zoo, which was really large and naturalistic as zoos go, but they would never let me in. After I had been at the job long enough to get my bearings, my longing for trees and animals came back acutely, and I remembered my dream of spending the day at the zoo. I went, and it changed the rest of my life.

Part Two

The SONGS *of the* GORILLA NATION

CHAPTER 5

Music Behind the Looking Glass Cage

The first day I packed a lunch and went to the zoo alone, I felt suddenly liberated. I was able to make a choice about where I went and how long I was there. I could choose to get away from the dirt and noise and smell of the city and escape to a place within me and around me without doom and danger.

This may sound strange to people accustomed to making active choices, but I had lived my life up to this point utterly unaware that I could choose a course of action and allow that choice to nurture me. Life was something that was beyond my control: I was always caught up in a swirl of sound, color, smell, and feeling that whipped my mind forward and then my body followed. People and events were something to survive, during which there was no rest to plan, to meditate, to have opinions or desires about the future. Going to the zoo was such a simple act, one among thousands that so many people take for granted, and yet this one act of decision quickly shaped the form and structure and spirit of my life.

I remember how nervous I felt as I approached the ticket stand and had to count out my money and interact with the cashier. I kept telling myself that if I made it past this point, the whole day and all its teeth and hot fur and greenery would be mine like a banquet that growled and grew. I knew the zoo held secrets for me to discover.

I was pushed forward by the restless line behind me and was finally released, floating slowly, like a wild river resting at last in a calm sea. It was raining, and the heavy droplets splattered the outstretched hands of the oaks and the birches, trickling down to the smaller waiting mouths of the berry bushes and the grass here and there among them. Ravens called and welcomed me, giving me all the news I had missed while I was away from myself. I felt born.

I went to see the lions and the kangaroos, the bats, the penguins. I saw horses and hippos, meerkats and mountain lions, giraffes, buffalo, and elephants. I took my time and let my tired heart decide the way, aware of a certain sense of growing expectation. Around a bend and under some low trees where the trail dipped, the pathway suddenly widened and washed into a grotto.

There was a bench there, wet, where water winding from the ceiling had slowly dripped upon it. I sat on the wet bench and breathed in the green, felt the cool, cleansing rain. I closed my eyes and listened to the quiet sound of drops and rivulets. There were no people, and yet . . . I cracked my eyes open, and through windows of glass and mist I saw them. Black and solid and timeless against the running and changing wet sat gorillas. Through the rain and a lifetime of waiting, they did not look at me, but they knew I was there. I sat still. I sat still. I sat for an hour, two, and three. I sat still.

They didn't look at one another, and they didn't look at me. Instead, they looked at *everything.* They were so subtle and steady that I felt like I was watching people for the first time in my whole life, really watching them, free from acting, free from the oppression that comes with brash and bold sound, the blinding stares and uncomfortable closeness that mark the talk of human people. In contrast, these captive people spoke softly, their bodies poetic, their faces and dance poetic, spinning conversations out of the moisture and perfume, out of the ground and out of the past. They were like me.

They didn't have to narrow their vision and cut the world apart. To look closely would have kept them from seeing and choked off the moving and breathing parts of the world, making it flat—worth little. And so we spent that first day, looking without looking, understanding without speaking. Writing poetry in our living, and reading it like weaving.

After that day I came back to the zoo at least once a week and sat with the gorillas for silent hours at a stretch. I would brave the bus ride and the ticket booth, knowing that a rich reward awaited me after the tension.

I never told anyone that I was going to the zoo; it was my secret world. And for a long time it was only about secrets and sitting. About feeling safe and calm. It was as if I was going there to take deep breaths, each a week apart and deeper each time, taking in and letting out in a rhythm that was as slow and ancient as time itself. I sat on the little wooden bench inside the grotto that connected the outside world with the shelter of the gorillas.

I began to understand the seasons of this place: the comings and goings of birds and insects, the tiny trails of slug mucus, and the night footprints telling a story on the parchment of the red dust. I began to understand the cold of the gray sky and the heat of the buzzing bees and honeyed flowers. A silence came to live here when the walking and staring human people stayed away, giving this place over to older things. I looked at the evidence of life as if I were walking trails with my eyes through wild book and paper, weaving understanding between words written in the dust, words written in eyes; making paths between glass and gorilla, reclaiming my home among persons.

I breathed. I began to wake up. I stretched and yawned in the soul I had forgotten, and the waking took months and years. I awoke a little more each spring and found rest for my weariness each enfolding winter. I shed more of my multicolored armor each fall. I was lucid in the swarm of summer. I opened my eyes. I opened. I started noticing things. I noticed how alike the gorillas and I were. Sometimes I was filled with joy, and I wanted to shout, "There are other real and gentle people among us! Human people are not alone! Look! Look!" Instead I stayed silent and I wrote about the gorillas—reams and reams—but ripped it all up and threw it away because it was all too private and I couldn't bear the thought of anyone seeing it.

As I wrote, and watched, and studied, in sadness, I grew to understand and identify with the gorillas, the way other human people did *not* identify with them; indeed, they treated them in ways similar to how I had been treated all my life. As I discovered and witnessed this, I often found myself awash in feelings that threatened to drown the tiny, emerging spark that was my hope for relating to humans in a new way.

"Look out, Cindy! Them things'll eat you if they get a chance!" a leering man says, stuffing a hot dog into his face. A little girl runs to his side, looking warily from behind a corner of his coat at the gorillas as they sit quietly. The man further scares the child, who is waiting to be wrong about her fear: "Them gorillas can kill and eat lions. I saw it in a book once." The girl files this away. Files herself away. She has heard the Truth. The man wipes his mouth with a napkin, wads it up, and lets it drop in the corner of the gorillas' house. The trash can is three feet away. "God, they're lazy!" he says. "Get up and do something!" he yells at the gorillas.

"Yeah! Get up and do something!" a teenage boy shouts. The gorillas just sit. They ignore, endure, remain. His shouts are deafening. They deafen. I am deaf. He and his friend then go to the glass and pound. "Hey!" They pound. They deafen.

They pound on the glass, on our bodies. I sit. I ignore, endure, remain. "*Stupid!* Man, you can't do anything with them. They're just stupid."

The gorillas don't speak human language, look the way humans look, move the way humans move. They are stupid.

This is why gorillas are captive. This is why crazy people are captive. We are the animals who don't speak the language, look the looks, move in the right ways. Captivity is for observing. Sit. Ignore. Endure. Remain.

It is easy for those who are not captive to forget that those who are remain individuals. An individual with a name, a family who needs them, a past that they stand on, and a future that they dream about. Maybe it is because so many people share the same past and dream the same dreams that they forget how lonely it can be to have a different past, a different dream. I knew what it was like to be in a prison. I knew because I was looking at it now with one foot outside the door, knowing the other would always remain inside. I wanted to know myself both as a person and as a primate, both as a free and as a trapped thing, both as a gorilla and as a human person wanting to know gorillas. I started to listen to what they were saying.

At home I read voraciously, trying to learn all I could about gorillas. Sometimes I read at the zoo and watched them at the same time. I was particularly interested in the fact that the apes we know today split from what would become the human line around five million years ago. This is a recent split, from an evolutionary point of view, and reveals a close and continuing connection between the human and ape nations. Yet in that short time the very success of the human animal, and perhaps the early evolutionary choices of the apes, have pushed existing ape populations down to paltry numbers and limited ranges. I wondered if the last of the gorillas would pass during my lifetime. This thought saddened me beyond the obvious tragedy; I had just found my people, and now they were going away. It would be my own extinction.

Conservation literature was often hard to read because it made me turn this question over and over in my mind. Often I awoke from terrible nightmares in which I was watching gorillas being killed, abducted, and sold. Sometimes I had to restrict myself to reading scientific treatises until the nightmares would lessen.

Through the scientific literature, I came to know that three races—or perhaps more accurately, tribes—of the Gorilla Nation have been identified. They apparently descended from an original single group that lived from Africa's west coast to the Rift Valley of West Central Africa. During the ice age their habitats shrank, forming islands of forest and isolating the populations that eventually became the three tribes that exist today. The Western Lowland gorilla inhabits West Central Africa, including Cameroon, Río Muni, Congo, the Central African Republic, and Gabon. Members of this tribe make up most of the population in captivity, and it was from this tribe that my gorilla family came.

The Eastern Lowland gorilla survives in eastern Democratic Republic of the Congo and near its border with Rwanda. The Mountain gorilla tribe populates the Virunga volcanoes of the Democratic Republic of the Congo, Rwanda, and Uganda. The differences in these tribes are in physical attributes such as size, hair color, density, length, and varying facial widths, and in culture as well.

Living, indigenous populations of Western Lowland gorillas number only 10,000 to 35,000. There are 550 Western Lowland gorillas living in zoos all over the world. Eastern Lowland gorillas are fewer, wild populations numbering only around 4,000. Only twenty-four live in zoos. Mountain gorillas are the most severely endangered. There are only 620 left in a 285-square-mile area. There are none in captivity.

Western Lowland gorillas, those who have not been

stolen from their homes, live in the warm, humid equatorial rain forest, gravitating to the rich understory where sunlight penetrates the canopy. Where logging and cultivation have occurred, gorillas use the secondary forest and may benefit, leading to increasing numbers. However, the poachers that arrive with the appearance of roads and the hungry loggers that make these inroads also kill everything they see, making the practice of pushing into gorilla habitat far more detrimental than beneficial. Open fields surround the gorillas' trails, and they routinely travel through them but prefer the cover of the rain forest and the zone at the edge of the forest. Here they eat shoots, berries, leaves, stalks, nuts, moss, and sometimes termites and ants.

The most important thing I learned from these beginning studies was that many human researchers believe that gorillas have spiritual feelings and religious sentiments.* They have cultures that rise up in different ranges, based on food preferences and other pivotal activities such as food preparation. They draw in the dirt with sticks, and some captive gorillas paint. They seem to have a developed sense of aesthetics. They have individual color preferences and grow attached to certain items, carrying them along as they go about their daily travels.

One gorilla even captured a mouse and kept it as a pet,

---

*During a lazy afternoon some years later, I saw the gorillas rouse from an afternoon nap, silently meet in the middle of their enclosure, arrange themselves in a circle, stare at one another for several moments, then break and return to previous places and their slumber. I always believed that this was profoundly important but will never know what it meant.

carrying it in her inguinal pocket, a naturally occurring depression between a gorilla's upper thigh and stomach. Eventually she let it go. Other gorillas have captured insects and treated them in the same way.

I was impressed with the breadth and scope of their feelings, their cultural habits, their problem-solving abilities, and their natural gentleness.

I was also surprised at all I learned about the worth and goodness of the Gorilla Nation. I had not been taught these things, but I knew what my culture thinks about gorillas. It has historically been difficult to raise sympathy for the gorillas' plight and combat the public's negative stereotypes about what gorillas are and what they actually do. A gorilla mythos lodged deep within the public's collective consciousness leads most people to fear and even loathe them without considering the facts. This myth was created even before whites had seen wild gorillas and was cemented when tales of first-hand encounters made it to the West.

Gorillas were called "black devils" by the white hunters of the nineteenth century, and popular portrayals of gorillas during that time (and often since) were built on the myth of the gorilla as a bloodthirsty man-killer: illustrations from this period show larger-than-life gorillas in ferocious postures, trying to tear innocent hunters limb from limb without any provocation whatsoever.

Contemporary advice to these hunters, however, takes advantage of the gorillas' often gentle curiosity. Hunters were told to patiently stand their ground until a gorilla came near, ready to learn and explore through close contact with humans, and brought the gun barrel into its own mouth in order to feel what it was. Then a hunter could pull the trigger.

Baby gorillas captured as a result of such human "bravery" were crated and carried out of the bush, never to return. When they arrived at their destinations (those that lived through the ordeal), they were usually dehydrated and suffering from serious infections, malnourished, with constant diarrhea and covered with sores. For every baby gorilla who lived to be displayed, nine times that number died during these early attempts to show them in zoos.

The Archaeology of Dreams

As I read about them in all the varied sources that I could find, my compassion for the gorillas grew, and I became invested in their fate. My memories of wanting to be an anthropologist came flooding back, and with a fervor and sense of direction I had never had before, I set out to find ways to accomplish my goal. It was very difficult and scary: I had abandoned school and was thrown into terror every time I considered going back. The very thing that urged me forward seemed to increase my fear; I had found peace with the gorillas and wanted to make them my life, but to do so I would have to give up the exclusive shelter they offered and work with people toward compromise.

I went to the library and talked to universities. Could I learn without having to go onto campus? Could I, with vision and insight, with a love of learning, find a way to visit the islands of others, to exchange ideas and benefit from their experiences? Could I at the same time turn away from the burning light of interaction and seek the quiet to make the experience part of me?

I was lucky. People helped me. I made many calls and followed many leads. I found an animal sciences program in a technical college that allowed me to work externally in mentoring situations. I learned that I could get involved in zoo programs and work with the people who ran them to expand my knowledge. People, to my amazement, assured me that all I had to do was *ask*, and they would help me learn and achieve my goals. Though I would still have to work with people, they would at least be people that I had begun to know: gorilla keepers, people within the zoo administration, veterinarians, and technicians.

When the people in the zoo's animal health department learned what I was trying to do, they offered me a position, and everyone I worked with was patient and tried hard to help me learn through social interaction. They also suggested reading lists that I could complete toward a specialization in animal behavior and a degree.

I was surprised to look around and realize that I had communicated with people toward a common goal and that they understood me. I still looked away a lot. I rocked and cleared my throat loudly and compulsively. I laughed at things that weren't funny to them, and I interjected thoughts that they didn't think were relevant. In spite of these things, they understood not only what I was trying to say in real social situations but what I wanted to do with my life.

I learned what was expected of me as a new college student, and I completed college coursework while I worked on site at the zoo. I was free to come and go as I pleased, and in addition to my formal learning I continued to read all I could about gorillas and sit with my friends for hours every week. I was never so happy. I had to deal with only a few people, usually one on one, and I could spend all my time learning.

In addition to sitting with the gorillas, which was my purpose, I was delighted to work with other animals in the capacity of my job. I helped raise a baby snow leopard, I assisted with a vasectomy on an orang utan, I was present for an operation on a wounded leopard, I helped with a giraffe that had an ankle injury. I accompanied vets to hands-on annual exams of hippos, colobus monkeys, and wallabies. I helped with hyenas that needed dental work, wolves that had tumors, and woolly monkeys that were suffering from age. I assisted in a study that focused on breeding rare red wolves. As an attendant, I looked after rheas, tapirs, cavies, and llamas.

I shared my salad each day with a mandrill that was in the hospital for chewing her wrist down to the bone. I felt we had a lot in common, and I would spend each lunch with her in the long, dark hall of the ward. I would turn and let her groom my head, then I would do the same for her. I would sit leaning against her cage and hand her choice bits of lettuce, carrots, and celery. I had been warned to stay away from her because she was aggressive, but I found it impossible to concede to authority on the matter and continued to sit with her until she was sold and transported.

In the evenings I was in charge of the family farm area. I let the pig play in the hose, scratched the sheep, and closed up the barn to let the rabbits run around and get exercise. I fed the raccoon, alone in its exhibit, by hand each night and sang to it. I would check the honeybee display to see that all was well; I loved to press my forehead against the Plexiglas of the hive and feel my head vibrate with a thousand humming, dancing bees. I felt like I was really helping and doing something I loved. I loved the animals and loved being alone with them.

As bright as this seemed, I did at times have trouble with zoo staff. I was odd and made some people uncomfortable. I had trouble following sequential directions, which was a problem because I was often left alone to complete complicated tasks. Most of my tasks were written down, and I did fairly well if I could do them in order. If something came up in the middle, which it often did, I would be thrown and unable to complete the remainder.

I was horrified to learn that I had left a pregnant goat with no bedding one evening and that she had spent the night, near her delivery date, standing on the concrete walking in circles. A zoo visitor had come to me to ask me a question about a sick rabbit while I was on my way to give the goat bedding, and my mind simply seized. I was thrown by this unexpected event. Between my nervousness about interacting with a stranger and the interruption of my careful schedule, I had simply not come back to remember this important task.

Another time I lost a master key to the zoo and was threatened with termination, as this was one of the most serious mistakes an employee could make. The key was found in the sheep yard by my co-workers after a hands-and-knees search.

The fact that I excelled at certain tasks—keeping records,

making keen observations, descriptively communicating information, and memorizing events perfectly—not only saved me but deposited me exactly where I wanted to be.

Because of these skills I was eventually assigned to observe the gorillas during one gorilla's serious illness. The staff knew that I had an affinity with the gorillas and that I might see things that another observer would miss. Perhaps they also knew by this time that I did much better when I was by myself, and being the kind people they were, they were simply trying to find a way to keep me busy and out of their way.

I believe they had started to care for me because I had begun to find ways to make it known to them that I cared for them and had good intentions. I cared about them as people, and after years of watching gorillas, I was learning how to tell them that. I knew to smile at them when I saw them. I knew how to put them at ease by sitting near them and to show interest in their lives by asking questions. I learned to show them a face that demonstrated sincerity and concern: I would consciously knit my eyebrows together and nod as they told me of their troubles. Like a gorilla, I would touch their shoulder to show them I was with them through their sadness and worry. When they told me something about their good fortune or relayed a funny story, I consciously laughed and brightened my face, squinting my eyes and turning my mouth up.

Because I cared, I wanted to do a good job in everything they asked, especially any job relating to gorillas, as I cared about them more than anything.

Determined to do a perfect job and prove myself, I spent hours in front of the window, immobile, watching every tiny movement the gorillas made. I meticulously recorded everything I saw. I was able to capture details others would have not seen. Most observers have to stop often to look down at the data sheets and write their observations, but I was able to write without looking away from the gorillas, because I could accurately envision the form I was writing on. This report, and its fastidious accuracy, led to my next big break.

My observation report was so thorough, precise, and insightful that the director of research* at the zoo met with me to ask me about my plans for the future. I stood in his small, cluttered office, taking in the specimens of animal teeth and bones, the pictures of trips far away, the books in "academese" that pushed and fell in on my attention from every corner. I was afraid to sit down and abandoned the idea entirely when I saw I would have to move a stack of research papers from the only unoccupied chair in his office.

He hunkered over my report, his tiny desk lamp throwing a pyramid of light over my work and glancing off his receding hairline. As he read over my notes, he chuckled to himself. I shifted from foot to foot. With a wry smile that was difficult for me to interpret, he said, "This is very good. Yes, very, very good." He leaned back in his squeaking chair to look at me through the slits of his eyelids. "You know gorillas pretty well, don't you?" Only later did I realize what a magnificent compliment this was. He asked me what I would like to do.

I told him about the program I was currently enrolled in

*The director of research at that time was William Karesh, the now-famous globe-trotting researcher.

and my interest in the evolution of form and culture. I told him that when my animal sciences program was over, I would like to pursue a B.A. in anthropology. I was anxious thinking about this, though, as I knew I wasn't equipped to go through the traditional protocol of regular college. The problems I had in dealing with my mentors and co-workers at the zoo would be magnified in a traditional college setting. My skills in math and linear processes continued to be uneven; despite my progress, I still had trouble understanding my peers. I loathed the thought of being trapped in a room with noise and bright lights, unable to get free and run in the trees and grass, unable to be still and quiet with the gorillas. And something as simple as trying to navigate my way through a campus would be very difficult and make it nearly impossible to learn. I also had trouble telling people apart and dreaded the experience of swimming in a sea of strange faces and unpredictable bodies, trying to find my way around a campus that would never look the same to me twice.

The director of research was sympathetic. His voice was calm and reassuring. I understood why he was so good with animals.

He told me that, with my obvious intelligence and ability to learn without constant guidance, I should certainly be able to study on my own. What I would need was someone to help me when I really needed it and show me how to navigate the system. He agreed to sponsor a series of gorilla behavior research projects. For the next several years I worked with him, and other helpful people, and was able to apply these studies and my continued zoo involvement toward a degree. I had no way of knowing at the time that I would eventually earn a Ph.D., but something deep within me knew that I was now on a road that I had been trying to travel since I was born. I began to have a vision of what I wanted to do with my life:

The door is open . . .
I know this is so,
Because I heard its weary hinges,
telling of the pain,
The pain
That is felt
When a very old door is opened
in your heart,
And it is a wonder,
Many do not hear
The weary hinges creak,
When one has opened
The door.

Beginning to Wake

As I look back on opening doors and new joinings, I am reminded of the first time I was ever allowed to go close to the gorillas without glass between us. The gorillas always went indoors for the night, where they found their dinners and toys all hidden carefully in the hay of their separate night rooms. There they would sleep. Each of their rooms was attached to the keeper's office, where night-watchers checked on them through the bars that separated the office area from their cages. In the mornings their breakfasts were prepared in the kitchen part of the office while they all looked on, issuing soft grunts of contentment and anticipation, knowing that soon their morning treats would be slipped to them through the bars, and then they would be let out to start their day in the habitat.

It was on such a morning that I was invited to come into this private place. I walked down the wet, early-morning trail through the empty exhibits to the massive wooden door that barred admission to the inner sanctum of the gorillas' private world.

Upon reaching the door, I found no way of getting in: it was locked, there was no latch, and I knew that the hallway behind the closed door was too far away from the inner office for anyone to hear me if I knocked. I stood, feeling panic, not knowing what to do. I paused unmoving before this great door, feeling the all-too-common sensation of shutdown overtake my senses. It always happened when I was confronted with this kind of situation. I stared, not seeing anything. Finally, I spied a tiny black buzzer button high on the door and pressed it. I heard the high, metallic complaint of the buzzer far off in the office, and shortly afterward the rubber-booted shuffling sound of the keeper grew louder as she made her way to the door. She opened it with a clang and a deep groan of hinges and invited me to step in.

She was pleasant and friendly, and I found it was not so painful to look into her eyes, which were much like gorillas' eyes; a deep and calm brown invited me in, and her steady habit of gentle throat-clearing sounded very much like a gorilla reassurance rumble. She was easy to like and easy to be near. I was able to talk to her.

She took me down the long hall into the inner office, and my excitement became hard to bear. I reminded myself to be calm. I felt as though I were being led to see some wondrous things, like the bones of the first person or the first evidence of other life in the universe. It was holy. One must look upon such things with quiet whispers of awe, with bowed head and bent knee, with the delight of the heart that binds all humankind with the infinite in the tiny. This was a tiny experience, a simple experience, like meeting one's long-lost parents as an adult. So simple from the outside.

The office door swung open, and I entered an atmosphere that was as rare as it was new. It was hot and humid the way the jungle must be. The scent of gorillas—like hot rhubarb pie—and sweet hay for the dozing hung in the air and enfolded me, intoxicating and heavy. The fragrant scent of a hundred good things to eat accented the sweetness: a riot of color and scent lay about in the form of oranges, celery, carrots, peeled eggs cut in two, red bell peppers, spinach, cinnamon-spiced apples baked fresh in the oven. Gorilla dung, smelling like horse dung, lay as an undercurrent. The wet concrete of the walls and floors offered only the faintest smells of captivity.

The red filaments of the heater overhead burned small dust particles that then smelled like the past, reminding me that this was an old scene, one that had turned each day to make pages in the book of the gorillas' lives.

I heard the small rustlings of hay and bodies and turned to look past the bars into the deep amber eyes of a family of waiting, watching gorillas. In turn they grunted deep rumbles that I felt in my chest rather than within my ears.

They regarded me. I was too full and floating to regard

them in return; instead I stood, transfixed and overflowing, feeling honored to be regarded and filled with their ancient gazes.

This was the first time I had met the gorillas. It was the first of regular visits I made after I became involved in their daily lives. But I never lost the sense of awe that I had that first day, and every time I entered the office, I closed my eyes to drink up the deep and wonderful spring that flowed into my senses. I never tired of cutting their vegetables, of scooping their dung, of carefully spreading the perfumed timothy grass around their rooms and arranging it into inviting nests up on the platforms in their rooms. Often I would lie in the nests I built for them and close my eyes to ride on the cloud of smells, sights, sounds, and feelings around me. I felt secure. I felt like nothing better could happen to me.

I heard the clang and moan of the opening door down the hall and woke up from the dream I was in. It was a dream about waking up, and it was coming true.

CHAPTER 6

The Difficulty of Lyrics Long and Unfamiliar

Over the next years I spent most of my time at the zoo, coming to know the gorillas and, in turn, myself.* Through my many research projects on them and their daily care, I got to know the gorillas as individual people and began to see basic patterns of humanlike behavior and discernible personalities, first on one side of the glass, then on the other.

There were two gorilla families at the zoo, and Congo, the gorilla I had first touched that

*In the fall of 1989 I was in a coma for several days as the result of a near-fatal asthma attack I suffered on a trip home to Illinois. It took seven months for me to recover sufficiently to live on my own again, and during this period I did not spend much time at the zoo. The fortunate result of this coma and recovery experience was that I was able to move into a low-cost apartment building for the disabled. As I gained my strength, I was able to stay in the building and take over a weekend attendant position on site and retain my apartment. This arrangement allowed me to finish school and to afford to live over the next several years.

wonderful day, was the leader of one of them. Congo was a silverback male who came late in life to Woodland Park Zoo. He had a kind and peaceful face and liked people, reaching out to them in ways that made them feel special. He would give them small gifts of his food or pieces of hay and try to coax them close so that he could have conversations with them. He softened hard people and made sensitive people come alive; he seemed to understand the deepest parts of people and communicate his acceptance of these parts to them. I admired his ability to do this, and I was touched that despite my social difficulties he grew to love me and accept me and even have a special fondness for me. His social gifts were especially poignant given the fact of his hard life—he spent years in a tiny concrete cage with minimal company—and it was a great pleasure to give him a better one and reward his great spirit.

It was after spending thirty-some years in tiny concrete cages at another zoo that Congo first experienced his spacious and verdant home in Seattle. His long quarantine after his initial arrival at the zoo was rewarded when the mechanical steel doors slowly lifted to reveal a new and boundless world of green beyond his indoor room. He peeked around the doorsill, only to quickly back away, as if the habitat beyond were an alien landscape or a world on fire. Then perhaps his babyhood memories slowly rose to the surface and some kind of vague recognition occurred. Finally, he moved his body around to sit in the doorway, the sunlight falling on his body. After a long time he moved out and sat on the one tiny area of concrete in the habitat, right in front of the door. It was a two-by-two-foot square of sad familiarity. He stared at the grass. He put one hand out and moved it over the top of the grass with a delicate grace. He brought it to his nose and sniffed it tentatively. Then he stepped out onto the earth for the first time since he was an infant. He grunted. He sat down. He let the earth reclaim his tired body.

Perhaps it didn't occur to him to move any farther than the small area he had become used to in his cage of concrete and steel, for it was days before he explored all of the grass, the hills, the stream, the trees, and the places to hide and be alone. Being alone was not what he liked best, however, and soon, to his delight, he had company: Jumoke, the daughter of gorillas from the other family, and Amanda, another female from a Canadian zoo.

Jumoke was the elder daughter of Pete and Binti, a gorilla woman who at one time lived with Pete and Nina's group. Binti was aloof and detached, and so was Jumoke, who grew to adulthood elaborating on her mother's tenacity and strong will. She had a sharp intelligence and always came at problems from unexpected directions. When Congo first came to the zoo, she quickly became aware of his presence on the other side of the fence and set out to solve the problem of their separation. She tried many ways to get closer to him, but when she was formally introduced and then allowed to join him permanently, her problems were solved for her.

Amanda was an unusual gorilla in that she had many male characteristics: her back was silver, her head shape was more masculine than that of other female gorillas, and her build was more muscular and broader.

After her own long quarantine, when she was let out into the yard with Congo, she refused to crouch and submit to him. Instead, as he tried with increasing confusion and frustration to let her know he was in charge, she attacked him and bit him back. At one point he stood heaving and wide-eyed after several running passes at Amanda in his best efforts to intimidate her; he looked over at me with a bewildered expression that showed that he really did not know what to do next. Eventually, the initial introduction was deemed a failure, and a long process of reintroduction was undertaken. At last, and in spite of her ferocity, it was successful.

Other aspects of Amanda's behavior were also more "masculine," and she approached both Jumoke and Congo in sexual ways, seeming to have little awareness of many dictates of gender protocol. She seemed to be guided more by inner motivations than by external cues in many areas of her life, much as I was.

In the other family the responsibilities of the silverback for many years fell to a wise and wide-faced gorilla named Kiki. He had been wild-caught in western Africa when he was around six months old. Like most wild-caught gorillas, he likely saw his family die in his defense and his mother shot from under him, but he was a forgiving and placid gorilla, only rousing to anger when it was clear that his intervention was needed to return peace to his group. He often sat, knees drawn up and breathing slowly, with a faraway expression blowing gently over his features. He always looked to me as if he were remembering something far away and long ago, perhaps the sheltering jungle of his new days and the family that now existed only in the wind.

Pete was in the unusual position of "co-silverbacking" with Kiki. Because Kiki was sterile, Pete actually fathered the many offspring that the zoo was happy to accommodate. Unfortunately, whether because of his personality or his memory of circumstance, Pete was neither happy nor easygoing. Where Kiki wore a serene expression, Pete scowled and glowered and had short patience with those outside and inside the glass walls of his prison. Where Kiki would quell outbursts and angry displays among his family, Pete often started them. It wasn't that he was unkind; rather, he wore his feelings on his massive shoulders, and they were passionate and tempestuous, meant to run a forceful course. And run their course they did; often he seemed to stay angry with people, for real or imagined offenses, for a very long time. His temperament did not seem to be passed along to his many children, however, and though he was aloof, he was never rough with his children and was a good father.

Nina, who was the same age as Kiki and Pete, was the matriarch of the troop. Very short and very round, she always reminded me of my grandmother. Even their faces were similar, which gave Nina a certain grandmotherliness in my eyes, beyond the quality she obviously possessed as a good mother to her own children. She was stoic. She sat quite still most of the time and regarded the zoo-going public with a quiet detachment and circumspection that led many to call her wise. Her black face was small and wizened. Her eyes, the dark amber of the Gorilla Nation, looked out on the world with an intangible quality of deep understanding. She was slow to anger and was respected by all; even the males would sometimes relinquish their places to her when she came near and looked at them without looking. Unlike many other gorillas, though, she would sometimes look directly into the eyes of people who taunted her and fix them with a kind of stare I have not seen in any other person. She kept her dignity.

Pete and Nina's son Zuri grew from a skinny, playful gorilla to be a handsome and cocky young silverback over the years that I knew him. Like any young man eager to prove himself and become his own person, he picked fights with his father as he grew to nascent adulthood: throwing things, barking insults, feeling persecuted, sulking alone in the back of the habitat. He herded the females around the enclosure and demanded their attention and admiration. Rightly, he felt trapped and frustrated and must have dreamed of the day when he could finally be on his own and make his place in the world. His captivity, like that felt so keenly among humans his age, kept him distant from those around him and made it difficult for them to form a relationship with him, but on wonderful occasions he would forget himself and become playful and defenseless for a few priceless moments.

Alafia, the youngest of the gorilla children, was wild, comical, and free. She experimented often with new ways of doing things, with new ways of seeing the world. With joyful abandon she would play interactive games with me; silliness, simple rules, and much repetition were the foundation and goal of each of these sessions. It is hard to believe, looking back, that she would grow into the very picture of her mother when she started having children of her own.

Haunting Strains

The zoo had a yearly ritual that took place the Day After Halloween. Watching it taught me a great deal about the various members of my gorilla families and, on reflection, much about human people themselves. According to a custom, the staff who worked with the gorillas would bring in their carved Halloween pumpkins and leave them in the habitat for the gorillas, who loved to eat them but rarely got to because of their limited availability. Watching gorillas participate in this ritual was a wonderful way to observe their unique personalities.

When Pete stepped out of the night rooms to witness the orange and grinning bounty upon the land, he would let out a call to the rest of the family. Something about the call must have declared "Pumpkins!" as the rest of the family would scramble out the door, down the trail, and into the midst of the round, ripe treasures that were everywhere for the picking.

The gorillas all ran here and there, looking at each pumpkin, judging them: size, ripeness, and color all mattered to varying degrees to each gorilla. The carved faces were examined and sometimes mimicked; Alafia would grimace at the leering pumpkins, making mirrors of plant and animal.

Each gorilla had a pumpkin-collecting style of his or her own. Zuri was the most comical: he hurriedly put a pumpkin under each arm, then quickly stuffed one in his mouth. He would get steadily slower as he struggled to find places to secure his booty. Invariably, he would end up getting one between his knees and another two in each hand in addition to those under his arms and the one in his mouth. Thus laden, he would mince as quickly as possible to a secret place in the back of the habitat and gorge to his heart's content. The flaw in Zuri's approach was that he would always go too fast and lose a pumpkin, usually the one between his knees. I would laugh, sympathizing with his fear of loss and his mighty struggle to get the pumpkin back between his legs in spite of his ever-shifting burden.

Alafia and Jumoke both went for size. In Alafia's case this had more to do with her love of wearing pumpkins than with a desire to eat them. She would find the largest pumpkin and then bend down with her rump in the air to squeeze her head inside it, pull herself upright with a strain, and then lumber and stagger about with her hands stretched out in front of her, blindly bumping into walls and rocks and other gorillas. Either she didn't realize she could use the carved eyeholes to see her way, or else seeing simply wasn't part of the game. She would always lose her balance, and the pumpkin's weight would complete her surrender to gravity. After the crash she would struggle to a sitting position, lifting the pumpkin up with both hands to make sure everyone was watching, and then start the whole process again. Intermittently she would lift the pumpkin, bits of guts and seeds stuck to her face, to reassure herself that she had an audience.

For Jumoke, on the other hand, picking the largest possible pumpkin was a matter of reveling in a grand prize. She seemed to gloat over the conspicuous enormity of her choice, disregarding the fact that the larger the pumpkin, the less tasty it was likely to be. It seemed that in her view, appearances were everything. Securing the goods for show was the deal.

Nina's technique was more dignified than those of the younger gorillas. As if she were in a supermarket, she looked over the offerings with a discerning eye at a skeptical distance. Employing her mysterious and infallible powers of assessment, she would linger over her choice. The smelling, tapping, and licking went on for agonizing minutes as the process ticked on and on. She would squint critically at a narrowing field of contestants and hold the grotesquely carved candidates in the air, turning them at angles, before choosing a few small, sweet finalists: the tastiest of the bunch.

Pete and Amanda seemed content to take whatever pumpkin happened to be close by. Whatever was easy was what they reached for.

Congo used none of these methods, as quantity never appealed to him, and neither did size or sweetness. Nor did taste, in the sensory implications of that term, enter into his judging. Rather than looking for tasty pumpkins, he was looking for a tasteful one: he liked the faces. With slow deliberation, he made his way through the pumpkin gallery, pausing here and there to delicately roll them over and appreciate their round visages. Finally, he would select the finest face, and standing up, he would hold it near to his own contemplatively. The pumpkin stared thoughtfully back at him as he ran his hand over its face and then softly over its low forehead. He appreciated this rare work of art. This beauty stood for something: he stood for beauty.

As I learned the rhythm of the gorillas' Day After Halloween, I came to understand the different styles of living that people had. I realized that human people have their own styles of gathering in the great pumpkin patch of life. Some get as much as they can as quickly as they can, even if they can't carry it all, and they take their booty to secret places where they don't have to share it. Some people seek out the prize that will make them look good, imagining that the sheer amount of what they have will make them the envy of others. And some people dance and play with the bounty of the earth and invite others to laugh when they fall. They are unafraid to search about blindly and stumble, they are unselfconscious when they have something unsightly stuck to their faces. These are the people who invite you to caper and trick merrily, lifting their jolly masks to see if you are watching and dancing along.

Then there are people who choose what is nearest, what is easiest. They may or may not enjoy what they happen upon, but the possibility of looking for choices does not appeal to them, or perhaps they are too tired to search. Others take care to see all that is available to them and choose the sweetnesses of life carefully and with great deliberation.

Still other people, like Congo, look for something more, something that can't be carried or shown off, tasted or reached for. They hold out for beauty and are open to the carving and the inner light of this life.

Syncope

In so many ways, large and small, I saw the best and worst of myself in the gorillas. But they had accomplished what I had not: the ability to remain open and communicate with others of their kind in ways that made them feel whole. The gorillas had an ability to empathize and to see value in others' desires for safety and happiness. This reciprocity of feeling and favor underpins the gorillas' relationships, as it does for those of humans. It is constantly communicated by those who will listen. I watched carefully and learned.

Gorilla people in captivity are forced—by living out of context, with other primates in charge of their needs and fates—to make the extra effort to communicate on our terms each day. I knew what it was like to long for silent communication, to need a private place to rest when the din would not cease and a sanctuary when the world would not conform to meaning.

It is no surprise then that I saw both sides of myself in the gorillas. We shared behaviors born both of our natural, archaic awareness as highly permeable beings, and also as reactions to the strain of a forced way of being.

I contemplated these ways of life often and thought about what I stood for. Physically, I thought about how standing up on two feet leaves you exposed. One's naked belly and chest and genitals are all uncovered and laid bare, as if standing has lifted a great warm cover made of the sacred space between body and ground. Like a plant uprooted, with the last of its anchor and succor falling in abandoning clods, we stretch up to the sky and let the close and nourishing earth fall away. This standing had often been too much for me to bear, and when it was, I would go and curl up somewhere, nursing the raw wound that my upright front had sustained in the million-year tearing away that my ancestors had undertaken. Stand we must, though, in order to move forward and reach up. So it often was for the gorillas.

Hiding the gorillas' food in high places beyond reach—in the bushes, trees, and walls—always led to them standing up to forage during the day. I had often seen Alafia stand up and carry tools from where she had found them to another place some distance away. Once she stood up and carried a long stick down to the pond—a distance of ten meters—to reach a candy wrapper floating there. I saw the gorillas carry armloads of food while walking on two legs, taking it back to the shelter and their nests to eat in peace.

Standing up to look over vegetation was also something the gorillas did—they would look over bushes to see where their family members had moved to. Such peering over bushes was a routine part of the hide-and-seek games that the young gorillas played.

Gorillas, both in the wild and in captivity, stand up to charge and to defend their families; they will pick up branches or fistfuls of vegetation as they swing themselves upright, then run, exposing their tender parts, shielded only by their ferocity and their love.

What I found I had always had with the gorillas was such vulnerability and ferocity and love. Our similarities went beyond perseveration, a need for space and a space for hiding; we were always drawing inward and exploding outward, sharing laughter out of fear and sharing a ferocious sense of justice, beyond mere caring. Our similarities also went beyond a difficulty dealing with the human race, sensitivities to the world around us and to the stereotyping in the face of the soullessness all around. Our affinity met in being filled with archaic darkness and persisting memories of a time when all things were one.

I felt somewhere deep inside that these gorilla people, a nation from the past, understood many things that we once knew but have now forgotten. They knew what passed next to them, and they knew what passed in the world. I wrote a poem about it.

The Songs of the Gorilla Nation

We look into the eyes of kin,
Both brought forth from ancient skin,
The songs of the gorilla nation
Are songs of where we've been,

We put them in a cage of glass,
And by their children
Children pass,
The songs of the gorilla nation,
Sung silent to the human mass,

In truth, to earth we both are tied,
They lived the truth until we lied,
The songs of the gorilla nation,
Were sung as nations died,

Only few now sing the songs,
And fewer still can sing along,
For the songs of the gorilla nation,
Are difficult and long,

But silently the old ones sing,
Behind their eyes the dry tears sting,
The songs of the gorilla nation,
Above the sadness ring,

Inside our hearts we hear the voice
of ancient souls within rejoice,

the songs of the gorilla nation,
To sing they have no choice,

Keeping time the raindrops fall,
The weary race can still recall
The songs of the gorilla nation,
When all sang one and one sang all,

For now the singers sing the last,
Their final children now have passed,
The songs of the gorilla nation,
Will only sing the past.

Singing the Future

I felt like the gorillas were living a great but subtle drama that was hidden to humans under a cold snow of unconsciousness. This piercing blindness made the gorillas want to hide. I understood hiding. It was the only way to keep from bleeding on the snow. And the gorillas did like to hide. They were lucky by zoo animal standards in that their enclosure was large and full of plants: there were huge trees throughout, and nettles, blackberries, and scrub, meadows of grass and copses of snowberry. A little stream ran through the habitat and ended in a small pond where ducks came to rest in the shade of logs at its banks. There were hills and cul-de-sacs. The gorillas would go behind the hills to sit or seek out the little caves in the underbrush. One gorilla, rather than leave the shelter of the public viewing area, would drape a burlap sack over her head and sit motionless with the world far away.

I understood the skill of hiding both in and out of plain sight. I remembered hiding under the hair dryer when I was younger, hiding in the honeysuckle thickets, hiding in the cave under the road, hiding in the closet. I was "off-view," in zoo lingo. Often now I wished I could sit in the midst of a crowding public with a burlap bag over my head. I watched with envy as the gorillas made nests in the sweet warm hay that was freshened for them daily.

I remember one day I sat watching from my familiar bench as Nina, recovering from a serious kidney infection, moved about the habitat. She had long had the habit of tearing apart the seams of burlap bags left in the habitat, and stretching the resulting long rectangle into a scarf, she would wrap it snugly around her shoulders. This probably helped her stay warm. She had been engaging in this behavior even more since her illness began. This unprecedented creative use of burlap hadn't been seen before in captive gorillas, and it gave insight as to how our early ancestors might have begun using materials available to them in these ways. For instance, Mountain gorillas sometimes drape small capes of moss around their necks, apparently enjoying the sensual activity; given their often cold, misty habitat, perhaps they too enjoy the feeling of having their own improvised scarves.

As I watched Nina this particular day, she paused to rearrange her burlap blanket upon her shoulders, securing it tightly around her when she saw that it was time to walk to the high back wall and receive the midday treats that the keeper always dropped to the family. She stopped often in her trek to the wall to arrange the burlap daintily so it hung evenly over her shoulders, which were still a little hunched and withered from the weight she had lost over the last weeks. She arrived at her usual place in front of the wall and began picking up dead leaves and other debris blown into the habitat. As the withered vegetation accumulated in her hands, I mused that she would have to be awfully hungry to consider these bits appealing in lieu of lunch. Certainly she must have had some other goal in mind. I really began to wonder, though, as she proceeded without any sign of stopping and with no fragment missed.

With one deft motion she hurled the debris in her hand downwind and watched with satisfaction as the breeze caught the many pieces and took them away in a winding motion.

The offending bits gone, she turned to inspect the area around her for any remaining bits that had escaped her efforts and eventually seemed assured that the spot was free of refuse. She stood up and with a flourish snapped the burlap from her neck, then let it billow out before her while she held two of its corners between her thumbs and fingers. After watching pieces of hay float away from it, Nina lowered the burlap, still billowing, to the ground. She spread it out on the grass and smoothed out the wrinkles. When the material was perfectly flat, Nina eased herself down in the middle of her picnic blanket and looked up to the keeper after letting out a long sigh.

She seemed suddenly self-conscious as she noticed the look on my face. It had been an incredibly involved set of steps, and we were awestruck. Determined to get lunch rolling, Nina acted as though nothing of note had happened and fixed the keeper with a look of riveted interest.

This incident helped me again to think about personal space and the ways humans establish it through external materials that are readily visible. I understood Nina's desire to mark off a space, to make it her own, to claim a small sanctuary in the swirling world.

I remembered all of the nests and shelters I had made as a child, the sticks and grass that I molded around me to give me peace, the juxtaposition of the material of nature and the artifice of design—a melting point between my flesh and the ground. In addition to Nina's "picnic blanket," the gorillas also made sheltering tapestries of limb and leaf. They would joyfully pile up the hay and cut saplings and then become intent on weaving them together in a perfect bowl, going around and around the inside, perfecting the structure with each circular pass like the sun around the earth. With complete abandon they would stop at some inner signal and flop down within the receiving arms of the fragrant bed and lie motionless until the late day called. This need for physical containment reminded me of my own. Containment silently reminded me of my physical boundaries—never solid and always in danger of disappearing—and kept me safe from the sensory onslaught of the outside world.

Watching Nina build her nests and throw her burlap about her, I gained a new sensitivity to the reasons that normal human people feel such a strong need for clothing and adornment. It is their way of creating safe space and defining a visible territory: a small one in a crowded world.

Of course, there are both aesthetic and pragmatic needs for barriers. Unfortunately for the gorillas, they had not much in the way of material to shield them from certain discomforts.

Like me, the gorillas did not like to get wet and muddy. They each had different strategies for coping with unwanted tactile sensations. When it was raining, some would run upright all the way from the releasing door to the shelter in the grotto, which was about a quarter-acre distance. Others would brave the discomfort of mud or dirt, racing to find a piece of burlap or a fistful of hay to wipe themselves clean, which they did with fervor. One gorilla would use a paper towel to wipe her face off and swipe at the crumbs left on her chest after lunch.

Noise was also a problem for the gorillas. In the summer especially, when sweaty crowds would press to the glass and the calling, yelling, and screaming would reach a fever pitch, the gorillas would become agitated and strike out at the public. The gorillas would bang the glass sharply, which, counterproductively, would bring collective squeals of fear and delight from the human people surging forward as the frenzy mounted in intensity. The gorillas and I would often retreat simultaneously.

Another thing I understood was the gorillas' social awkwardness around human people.* I did not seek other humans out, and it did not occur to me to share my deepest feelings with them. Like many of the gorillas, I was distrustful of them and most times found them incomprehensible. Rarely if ever had I felt an urge to secure the comfort of other people when

*Though socially awkward with most humans, the gorillas—like me—seemed more comfortable around women. They seemed to feel less threatened, less likely to need to hide or stiffen, than when men were near. Louis Leakey, the famous paleoanthropologist, knew what he was doing when he sent women like Jane Goodall, Dian Fossey, and Birute Galdikas into the field to learn from apes.

I felt close to breaking. As I grew to know the gorillas, however, I began to feel a growing need for the presence of another, someone with whom I could share the mundane and the heartbreaking. I needed to communicate with someone else, but I didn't know how.

During one particularly bad week, when everything had gone wrong, I felt an irresistible need to go to the gorillas and get help. I didn't know how to do it or, at that point, why I even needed it, but blindly, I found my way to the zoo and numbly stumbled down the trails to what I hoped would be waiting peace. I wound my way to the night cages, where the gorillas were awaiting their release. No sooner had I made it into the light of their presence than I felt my rigid body and the sharp parts of my heart begin to crumble. I fell to my knees and crawled with blind deliberation to where Congo was sitting on the other side of the imposed barrier between us. As I struggled to the window of the cage, I looked up into his huge warm face. He knew, probably even before I showed my own face, that I was falling apart. He saw in me what I could never see in others—he could read my emotions from my expression. He rushed over and searched my face intently. My vision blurred, and tears spilled out of my eyes and dropped onto my clothes, my boots, the floor, and the barrier between us. Still we looked at each other. Congo moved toward me, put his massive shoulder against the window, and motioned with his hand for me to lay my head there. I let my head fall softly on the place where his shoulder had been offered and cried silently. I could feel the long hairs of his shoulder brushing the side of my face. I felt sad. I felt guilty. Why should this person, taken from his murdered mother, abused as a baby, living in a prison of my own kind's making, care about my pain? I realized then that it was what he was born to do and what he did. It was the foundation of his inner dignity to care. In that moment I started to understand human men in a way that I never had before, and my fear of them began to lessen. I began to see that the core of my being was a great deal like this male core: looking on from the outside, blank-faced, with a deep and abiding need to protect and comfort in a world where my ways of feeling and acting no longer had context. My archaic animal nature had no place in a modern world. My kinds of sacrifice were no longer needed in a world of buildings and machines.

Congo, a man of sacrifice and ferocity, showing his care and his invisible strength in a jail built by those he loved, inspired me to open up and extend my heart to the world around me. I would no longer allow the great permeability of my spirit to lead me to seek smaller and smaller shelters; I would let myself bleed out into the world and let it into me. I would *be* among people, no matter what the pain. And though it *was* painful, though it meant fighting a losing battle for the gorillas, though it meant that my way of being a scientist would be rejected by most, I would turn my prison into a temple. I would reach out, I would share my pain and the pain of others.

We are all in cages of one kind or another, I suddenly realized. The dignity of caring and bravely showing it was something nobody could take away. It is a waiting possibility in all of us. It lives in the silently crying people, the people going about their daily lives in the prison outside the one we built for gorilla persons, and it lives in them also. As I thought about these things, my face still buried in Congo's shoulder, the minutes passed slowly. He softly grunted his reassurance.

After that day I found that I was very sad watching Congo, Pete, and Zuri threatening the noisy crowds, throwing sticks at the public as they jostled at the window. They would charge with grimacing faces toward the unrelenting stream of maddening humanity, unable to protect their families from the unconscious human people creating a din. These ineffectual attempts to intervene on their families' behalf and fulfill their manly duties would always draw loud shrieks, nervous smiles, and staccato bursts of laughter from the crowds. Each of these behaviors was threatening to the gorillas and, in a vicious cycle, led to more displays, more laughter, more teeth on both sides. The audible sounds of this broken carnival equaled the empty howl of soulful discord, ricocheting off the walls of one's ears, one's body, one's soul. The men on either side of the glass mirrored what had gone wrong in the world, their echoes of each other ringing in the ears of living things both far and near—a threat that grows within and shouts itself in colors and paints everything. It rings and sticks. I think that it is this insidious noise that men of our species rage against to no avail and that women fear and wish men could make go away. It is there in our shouts and also in our most quiet of whispers. It is the wind in our language.

I had many more chances to learn about sharing and communication from the gorillas. Koko, a gorilla who has been learning American Sign Language since 1972, has a reported vocabulary of around a thousand words and arranges them into sentences. She has been able to learn some written English and recognizes many words, including her name. She also shows an amazing understanding of spoken English. If a gorilla can learn human language, why couldn't I learn theirs?

My understanding was simple at first, like someone learning a second language or, more accurately, a small child learning a first. Male gorilla "hoots," much like frightened human calls, warn of possible danger or impending attack, and the deafening and unmistakable "wrraahs," like our screams, are roars that are intended to get something threatening to back off immediately and warn others that they should run and hide. The long, rumbling grunts or belches that gorillas make in their throats when they are happy and content are much like our own sighs or grunts of satisfaction. Gorillas use these contentment grunts to reassure one another that everything is all right. Gorillas cry when they are sad, and their crying sounds like human crying. One person claims to have seen a gorilla cry real tears after being taken away from her family by poachers. Gorillas laugh and smile, frown and show surprise, express affection and frustration. Head-bobbing is a threat among gorillas, crouching down or lowering the head or the whole body is a means of avoiding confrontation, and pats, hugs, and kisses are offered for reassurance. Chest-beating—which can actually be done on the chest, stomach, or thighs using cupped hands—is a complex and contextual signal and can express agitation, joy, exuberance, an invitation to play, or dominance. These were words I understood.

I remember a key incident when I made a real breakthrough in terms of my understanding. I was out on my small wooden bench taking notes about the things the gorillas were doing, when it came time for their midday snack. I saw the keeper picking her way cautiously along the narrow trail that ran across the crest of the enclosure wall; her metal bucket, brimming with tantalizing fare, was swinging at her side. In spite of the chill and the rain, the gorillas made their way to their customary spots: Jumoke close to the granite wall almost directly below the keeper, Amanda off to the left settling in on the hillside, and Congo sitting back farther away in a position of proximity that suited his dignity. They caught the treats thrown to them one at a time, always starting with Congo. Baked apples sprinkled with cinnamon left their toothsome traces in the air, while sliced oranges added fresh overtones to the perfume of the repast. Shredded Wheat biscuits, celery, and peanuts all came down into the hands of the waiting gorillas with careful aim to each of them in turn, a shower of arcing bits that were quickly popped into waiting mouths.

When the bucket was empty, the keeper set it down and watched the last of the treats disappear. Before long Congo came closer to her and sat down once more. Fixing the keeper with an intense stare, he raised his arm toward her and then moved his outstretched hand repeatedly in a curling motion that clearly meant "give me." The keeper smiled as she reached deep into the pocket of her old tattered coat for the prize Congo knew was hidden there.

The keeper produced a soft-boiled egg and deftly tossed it to the waiting silverback; he caught it with both hands, brought it to his mouth, cracked it with his teeth, then peeled it using only his lips. As he bit, a tiny trickle of soft yellow yolk spilled from his lip and ran down his chin as he closed his eyes. When he had swallowed the thoroughly relished egg, he grunted his approval and clapped his hands together to punctuate the end of this delightful interlude.

Noticing that the keeper was still on the wall, Congo, a hopeful look on his face, repeated his asking gesture, but she turned her coat pockets out and lifted her palms up with a shrug to prove there were no more eggs. With a disgusted look, he waved a fist at her and turned his back to her place on the wall.

Waving to me, she gathered up her bucket and once again made her way down the trail, seemingly unaffected by Congo's disgust. Her bucket swung and clanged hollowly as she disappeared from view.

The gorillas came back into the grotto one by one, and each of them came by my place at the window to peer into my face and look at my own lunch. Jumoke was first, putting her face close to the window and examining the newly produced contents of my backpack. Amanda sauntered over to take in the items while picking at the hay in front of her, pretending not to be interested. Nothing I had seemed to capture her fancy.

Congo came to greet me last. His brows were furrowed as he sat directly in front of me, plopping his great bulk down so that he was turned slightly away from me. Then he made the asking gesture to me. I sat stupidly, suddenly feeling a little like I did with most people in the human world. I didn't know what to say or do that would be appropriate. I was surprised to feel embarrassed. He made the gesture again. I held each individual item from my lunch and lifted it, along with my eyebrows, to ask which food he was interested in. For the first time it occurred to me to ask for clarification when I did not understand or know what to do. Always in the past, embarrassed by not knowing what to do and overwhelmed with the presence of another person, I would walk away or act brashly, taking control and imposing my own answers on the situation. That didn't work with gorillas.

Eventually I showed him the entire contents of my lunch: leftover macaroni and cheese, different fruits and vegetables, and a bottle of Gatorade.

He pointed to the bottle. Still feeling stupid, I shoved the bottle against the window and shrugged my shoulders—it wouldn't fit through, I tried to say. He pointed to the wall where the keeper threw his treats. He knew the trail led to a secret area close to where I sat. He raised his eyebrows. *"Walk up there and throw it down to me . . . what kind of stupid gorilla are you, anyway?"* he seemed to say.

I shook my head and pointed to my seat and notes, in a feeble attempt to demonstrate my duties. He turned his body away from me and reached back to bang the window with his fist, pursed his lips, let out a raspberry, and then pointedly ignored me. Occasionally he would turn to look over his shoulder and purse his lips in my direction. He didn't need to say it in English; I knew what was going through his mind.

This was one of the first times I remember knowing for certain what another person was thinking and feeling, and that my actions were a direct cause of their subjective experience. Something about the directness of his communication, combined with the honesty of his body language and his emotions, painted a kind of consistent and forthright picture that allowed for a moment of communication that was, paradoxically, more intense and more subtle than that of a human person. It demanded that I stay engaged until the moment had resolved with both of us as participators. It is clear to me that not only do apes have a language that is complex and holistic, but by communicating with us, they illustrate that it may be we who are less skilled at the art of sharing true subjective experience.

This lesson was particularly useful to me as I was working my way through school at the time. Because of my earlier studies, and because I knew I would have difficulty in a formal classroom setting, I made sure I was very prepared for all my coursework before I ever took a formal class. This meant that all I had to focus on when I finally attended a class was being considerate of other people and learning how groups worked, which became easier as I studied and spent time with the gorillas. As I learned their language, I began to have a context for human communication that made it meaningful to me in a whole new way.

Very cautiously, I tried to apply the things I'd learned from the gorillas in social situations. I tried to put people at ease by acknowledging them with quick sideways glances and smiles—which evolved from submissive primate grimaces and are intended to convey that no harm is meant. I learned how to avoid arguments by putting my hands up with the palms facing outward and head down. I asserted myself by making myself bigger, putting my hands on my hips and pushing my elbows forward, with my brows knitted and jaws tense. I learned to take turns receiving and sending signals to those I was interacting with. A feedback loop took shape: I felt generally more calm in my life because I spent so much time with the gorillas; because they made me feel calm, I was able to watch and learn from them. By applying the bodily and verbal language components I had learned from the gorillas, I was beginning to have more social success; this led to less tension for me when I was in social situations, and that in turn enabled me to relax and read people better. This process allowed me to return to the gorillas, knowing more about how human society worked, and learn about the feelings and motivations that underpinned the gorillas' actions and patterns as individuals and as a group. I learned on a new level that communication is meant to convey and evoke *visceral* feelings, not just rational or mental feelings. Though I had understood what fear and anxiety felt like on a gut level, I now began to understand other, more complex emotions.*

I was able to write a poem from two points of view. I had never done this before.

I had to laugh,
And the way I looked at you,
you had to laugh at me,
And you . . .
You laughed 'til tears
Rolled down your laughing face,
Then you hugged me.
We knew it was good to laugh.

*Captive gorillas and other primates also feel fear and anxiety, some chronically. This was expressed in a variety of ways, from spinning and self-injury (already mentioned) to hair plucking. Amanda had the curious habit, recorded in other captive gorillas, of vomiting into her hand and reingesting the material. She would do this in a compulsive way, again and again for hours.

Anger

As I learned more about communication and connecting, I began to understand emotions that had previously been only abstractions to me. I learned that anger could often be about embarrassment and diminishment. When Zuri, as a young gorilla man having no proper objects for his budding feelings, developed a strong attachment to a human female volunteer, he showed off in front of her and was always watching to see if she was paying attention. He did his best to impress her with his dignity and self-control. One day while she was feeding him the midday snack from the wall, he failed to catch a Shredded Wheat biscuit, and it hit him on the top of the head. The biscuit exploded in a rain of crumbs that showered down across his chest and shoulders. He was angry. He wouldn't look at her for weeks but rather wore a purse-lipped frown when she was near. I realized that anger and grudges of this sort are a product of people feeling embarrassed, and I started to work hard not to make people feel this way.

I also saw a different kind of anger demonstrated—anger caused by another's selfishness or thoughtlessness. Kiki would become angry when one of the young gorillas hurt another one. He would get angry if one took a treasured morsel from another. He had a refined sense of justice, which made me appreciate better this kind of anger on a gut level, both when I felt it for myself and when I must have made others feel it. The idea of sacrifice for the greater good and the capacity for regret are central to this kind of justice, and in many instances the gorillas showed a grasp of both impulses.

Once Zuri, in the throes of a full-blown display, flung every stick in sight, then searched for another projectile—in vain. So he grabbed Alafia by the back leg and raced away with her, scraping and bumping her over the ground; she was screaming as if her life was over. Nina and Pete immediately confronted him, and he dropped the whimpering Alafia. She sat crying with an injured, wide-eyed expression on her face. After heaving to catch his breath and gaining his bearings, Zuri went and sat next to Alafia, put his arm around her, and peered into her eyes to apologize. I had no doubt that he regretted what he had done.

Concern

I saw that anger and caring could be flip sides of each other, and that this kind of regret must of necessity be informed by an ability to empathize and to see value in another person's desires for safety and happiness. It compels one to act fairly. This reciprocity of feeling and favor underpins gorillas' relationships as it does those of humans. It is constantly communicated to those who will listen.

One such example of this kind of caring came early in my years of working with gorillas, during a time when Nina was very ill with a serious kidney infection. She refused to move or to eat and was losing weight rapidly. Her family was understandably worried about her.

One morning she was barely moving at all, and both Pete and Zuri were becoming very upset about her behavior. Zuri put his fingers to his mother's rectum and then sniffed them, looking at her with a worried face. Pete sat closer to her than was his custom, eyeing her sideways. She continued to lie motionless, her arm draped over her eyes and her abdomen bunching in pain in regular cycles.

This went on for hours. Finally the keeper appeared on the top of the habitat wall with the midday snack, including many new items to entice Nina to eat: peanut butter sandwiches, Popsicles, mangoes, special fruits, and yogurt.

Pete and Zuri were no doubt hungry after watching over Nina and, on seeing the treats, headed enthusiastically over to the wall. Halfway to their regular places, they realized Nina was still lying in the grass and stopped, turning toward her. Torn, they stood in their tracks weighing their concern for Nina against the lure of the beckoning feast. Pete barked in frustration while scowling in her direction. Her eyes remained closed, and her face was knotted in pain.

Their lips pursed in consternation, Pete and Zuri strutted back with stiff gaits to where Nina lay. Once they were near her, the two males began flinging sticks around her immobile form. They would dig a stick from the hay, stand upright, then charge toward her, only to let the stick fly in her general direction as they fell to all fours and streaked past her with single barks.

Though their objective was to force Nina to rise and join in the group's meal, Pete and Zuri's charges jarred and frightened zoo-goers who were unaccustomed to such noise and force. They were openly outraged at the beastly treatment Nina seemed to be receiving from her family.

Around me I heard their comments. This behavior, some said, was evidence that gorillas are aggressive by nature, that they are just plain mean. When I tried to explain that Nina was sick and that concern for her was motivating their behavior, my ineptitude at communication left them free to pursue the idea among themselves that they were now witnessing the law of "survival of the fittest." They seemed to believe that Pete and Zuri's efforts served no other purpose than to weed Nina immediately from the gene pool.

"This lady says that the gorilla lying down is sick," one woman explained to her son. "Those other gorillas are trying to get rid of her because she is weak and useless to the group. Survival of the fittest—that's how nature works."

Feeling panic rising within me as people became swirls of color and noise, I tried to explain that her family was trying their best to get her up to eat and that the gorillas were really worried for her. Someone told me they had a strange way of showing it, then left abruptly. I was not explaining well enough and was sinking into a place of stony withdrawal. The last thing I heard was a comment that I was cold-hearted. I retreated from my attempt to interact with the people around me and, as I had so many times, turned the whole of my attention to the gorillas.

Unfortunately, the chaos whirling around the gorilla family was hardly more comforting. Just as I focused on the displaying males, one of them launched a thick stick that sailed through the air and struck Nina on the head, making her wince and duck. Gorillas rarely actually hit someone with objects they fling. Nina and I were equally surprised by this event, and in stunned silence she reached to her forehead and inspected her fingers, showing traces of blood. Pete and Zuri stood heaving and stiff, waiting to see what would happen.

With an expulsion of breath and a furrowing of her brow, Nina rolled over and got to her feet. She was weaving unsteadily. She started toward the wall. Pete and Zuri strutted closely behind her. She sat in front of the wall and reluctantly began to eat the treats the keeper threw down to her.

It was clear to me exactly what these men were doing. They were looking out for the health and well-being of a loved one. This kind of caring is something that male gorillas are known for.

So did the gorillas teach me to value the feelings of others. Most gorilla men feel a powerful need to help those in their families who need them and to give whatever they have. It tears them up if they cannot come to the aid of one who needs them.

Gorilla men are ever vigilant, looking for signs of danger in order to alert their group and instantaneously guide them from peril. As a result, their families are always deeply tuned in to the moods, demeanor, and body language of their silverback leaders. The men set the emotional tone for the entire group.

When their families are resting or sleeping, silverbacks post themselves as sentries at the group's edge and stay awake and alert, vocalizing and chest-beating deep into the night. They settle squabbles and come to the defense of young gorillas; they have also been known to adopt orphan babies, allowing them to travel by their side and even sleep with them in their night nests; the small and fragile orphans against the warm massive body of their protectors.

In the wild, gorilla men will free group members' hands or feet from poachers' wire snares by sliding their large canine teeth between the wire and the wrist or ankle until the snared gorilla is able to finally slide out. This skill has saved many lives, as gorillas who break free with the snare still cutting tightly into their flesh will eventually become crippled or will suffer from gangrene and ultimately die of its complications.

Silverbacks also slow the pace of their groups as they travel so that sick or injured members can keep up, sometimes even sleeping near them until they die, if their affliction is fatal. Gorilla men will also lead their groups back from necessary feeding forays to reassure dying members who can't go on.

I once saw Congo intervene on behalf of a human person on the other side of the glass. A little girl was being shoved by a group of boys, which caught his attention immediately. He pursed his lips and pounded repeatedly on the viewing glass until they left her alone. As the boys jumped to face Congo, their eyes were wide and their mouths were open. *Shit!* I could almost hear them thinking. I chuckled.

Humor

Incidents like this one taught me a lot about humor—what it is, where it comes from, and why things are funny. I often witnessed interactions like that between Congo and the boys, or between gorilla family members, and I would feel a little pang of fear or tension about it, then laugh. Humor, I came to understand, happens as a "relief response" to something that scares you but doesn't end up hurting or killing you. It could have hurt you, but it didn't. Maybe that's what people mean when they talk about the "cosmic joke."

One incident that taught me a lot about the way fear and humor are closely bound happened one great and sunny day when the gorillas were all in fine spirits and happily enjoying the atmosphere. Zuri was in a giddy mood. He ran and capered about, tagging his elders from behind and inviting his little sister to play. Comically, he tried several times to hide his massive 350-pound frame behind some straggly stalks of nettle in the back of the habitat. In good sport, Alafia intentionally ignored him as part of the game and pretended to be completely immersed in searching for bits of food in the yard. She would surreptitiously glance at her brother from under her brow as she moved toward him, her search purposely taking her within his pouncing range.

As she drew nearer, his muscled frame bunched in anticipation—then he leaped on her, with his canines fiercely exposed. He wrestled her to the ground and mercilessly dug his fingers into her neck and underarms until she was breathless from panting laughter. Her mouth open and her eyes squinting shut in a mask of delight, no sound would come out of her until she struggled free and drew a loud breath while trying to escape Zuri's hands.

Again and again this scene would play out. Alafia would make a daring escape from Zuri and go to sit very close to her father, who was watching this routine lazily. She would shoot Zuri looks that were meant to egg him on, as she knew Pete would protect her. He played along by occasionally putting a reassuring arm out and touching her shoulder. Then Zuri once again took his position behind the nettles, whereupon the whole cycle would start again.

Eventually Alafia, being the young and distractable gorilla

she was, tired of the game. As her grinning older brother trembled in anticipation behind his nettle stalks—which were now limp and falling at angles from ambush after ambush—she ignored him. It was really sort of pathetic to see Zuri still bunching for a "surprise attack" behind one or two bent and broken plants that did not conceal him in the least.

Alafia decided to reverse the game to make it more interesting. Still pretending to look for food in the grass, she edged to a large tree out in the habitat and quickly slipped behind it. She peeked around the trunk just long enough to make sure Zuri was still behind the nettles, then pressed her face tightly against the bark of the trunk away from his view. There she stayed. But she forgot that keeping an eye on your approaching quarry is absolutely critical, for when she peeked back around the trunk, the object of her plan had disappeared.

Zuri, a seasoned veteran of such games of deception, had taken the opportunity to circle around his sister while she wasn't looking. She stretched her short neck out farther around the tree with a puzzled look, scanning the habitat for her brother. Then she resumed her hiding place. Suddenly out of nowhere Zuri roared with all his might and grabbed for her with both hands outstretched in two gargantuan claws. Alafia's look of perplexity turned to terror.

Seeing a gorilla shocked and terrified makes one's whole body tighten in sympathy. But seeing a gorilla weighing 175 pounds leave the ground like a rocket went beyond funny—it was hysterical.

Alafia seemed suspended in the air. As she hovered at the arc of her trajectory, her arms and legs jerked out, and her tongue stuck out stiffly from her wide mouth. Her eyes bulged, and she made a squawking noise. By the time she landed, stark terror had given way to the realization that she was not in any real danger. For her, however, this realization did not lead to laughter—as it had for me looking on—it led to fury at her brother.

As soon as she had landed on the ground and got her tongue back in her mouth, she grabbed at Zuri, her teeth bared and her fists clenched. She bit and pummeled him, barking angrily. Zuri, his arms blocking his sister's wild assault, turned from the fray toward me with a bewildered look on his face as if to say, *"What? Wasn't that funny?"*

It *was* funny. It was funny because it was scary, and because everything turned out all right in the end. Drawing from this example, I became fascinated with making jokes that capitalized on people's fears, including my own, understanding that sharing fears is funny when you have reason to believe you will pull through together. Humor itself, I learned, can be a creative way to ensure that you do pull through.

Motivation

One day the gorillas were all inside the shelter, and I was huddled on my bench. Alafia was clearly bored and looking for something to do. She seemed restless as the older, calmer gorillas of her family ate pieces of food and meditated on life inside the habitat. Suddenly her attention was riveted; her body became tense, and all her attention was focused on the east wall of the grotto.

As I followed her gaze, I saw that there was a small white

moth high up on the wall; it was flitting around and then landing in the same place again and again. Alafia raced over to the rock face and tried to swat at the moth, but it proved to be too high for her to reach. She sat down before the wall and, without moving her eyes away, stared at the moth for a long time.

Finally, she got up and moved away from the wall. I thought she was probably bored and wanted to do something more than sit with her eyes glued to an insect she couldn't reach. She pushed aside the hay in the grotto as she moved here and there, looking for something. Finally she found a short, sturdy stick, about two feet in length, which she tucked under her arm with great deliberation. She continued to sift through the hay until she found a second longer, thinner stick. The procurement of these two sticks lasted a full minute, while a look of absolute concentration was spread on Alafia's face.

Returning to the wall, Alafia carefully lodged the shorter stick at an angle against the wall directly below the moth. Then she climbed up the shorter stick, using it like a ladder. Never letting the moth out of her sight, she swung the second stick at it, hitting it squarely on her first try.

Fluttering feebly, the moth fell to the ground and twitched on the hay below the wall. Alafia dropped her weapon and leaped from her ladder stick. She grabbed the wounded moth and popped it into her mouth. Mashing the moth around her mouth and over her tongue, she savored the rare treat.

I was amazed. I never heard of gorillas using two tools at once, but I had just witnessed a gorilla actually hunt with a weapon while using a second tool. This was quite an operation and required great attention; one might even say that it took perseverative thought. In order to carry out such a complex task, Alafia had had to approach the problem from several angles, weaving and picking up strands of thought as she went along, like mental macramé, and then carry her plans out one step at a time while not allowing any other input to distract her or sway her from her purpose.

Perhaps, I thought, my ability to concentrate so exclusively on my own thoughts was advantageous in divergent problem solving. If this was true, then it could be possible that such perseveration may have been a widespread trait in our ancient ancestors: the tool makers, the shelter builders, the weavers, and the sewers and planters. Perhaps my own way of being is a very old way of being, a gift from my forebears.

Religious Sentiment

Some of the eventualities of life are too upsetting and too serious to laugh about. I learned this when, in my enthusiasm for exploring humor dependent on fear, I went too far and my behavior was labeled inappropriate. After careful study, I learned that there are categories of fear that are best not joked about, especially those centering on irreversible changes that make people feel helpless. Many people, for instance, do not like to joke about death; they simply do not find it humorous. Others do not want to joke about getting old, or permanently losing things they love, or their children moving away. These things, after all, *can* hurt or kill you. This, I began to understand, is what religious ritual is for; helping people cope with things that scare them and are beyond humor.

One amazing ritual I witnessed began on a cold and rainy

day, a typical Northwest day with a sky of steel gray and the ground running with water as though the earth were crying profuse tears for want of the sun. The gorillas, unhappy about the rain, began coming out into the habitat. Pete stopped in the doorway and looked up to the sky, squinting against the drops that splashed off the sill into his face. He appraised the soggy mess outside with a disgruntled expression. Reluctantly, he put one of his great hands and then his foot outside onto the path, curling his lip as mud oozed up coldly between his fingers and toes. He slowly brought out his other foot and hand and planted them into the mire. His lips were now pursed tight, and he seemed to struggle with the necessity of letting go of the toasty rooms inside. Finally resolved, he walked stiffly to the shelter, shaking his hands and feet after every few steps.

Zuri was right behind him, but rather than ease slowly into the sopping mess outside, he had a different strategy. He grabbed the sides of the door to the yard, gave several back-and-forth movements like a skier at the starting gate for the slalom, then exploded in a blur toward the shelter, running upright the entire way in order to keep his hands dry. Once inside he grabbed a handful of dry hay to wipe himself off, clearly disgusted with the rain and mud clinging to his body.

Nina was more stoic than either man. Without stopping either to regard the rain or to curse her fate, she plodded out the door and went directly to the shelter. Sometimes Jumoke would hang on to Nina's rump and walk upright so she too could keep her hands clean, but today Nina made her way to the shelter alone. She was far along in her pregnancy with Alafia, and her gigantic belly nearly brushed the ground as she plodded along on her short limbs. Jumoke was nowhere in sight. I wondered where she was.

I didn't have long to wonder. Without warning Jumoke came streaking out of the doorway and tore across the back of the yard, only to disappear behind the trees. The trees began to shake violently. Jumoke, emerging above the copse, had climbed one of the small trees and was forcing the top of it—to which she clung—to wave widely back and forth as she shook it with all her might.

I could hear her screaming and hooting into the falling rain, her wildness a strange and stark swatch of black color against the complacency of the steady shower. I grimaced as the little tree bent nearer the ground with each wild career.

Suddenly a crack that sounded like lightning turned out to be the top of the tree coming off—with Jumoke still attached. The underbrush received her as she flailed out and yelped, and then all was still. A few leaves, stripped from the trees in the frenzy, fluttered down with the rain. *She's dead,* I thought.

The gorillas and I looked at each other and then stared at the spot where Jumoke had disappeared. Endless moments went by, with no movement, no sound. Then, just as suddenly as she had fallen, she exploded from the underbrush, still holding on to the top of the tree and dragging it behind her. With her new weapon in tow, she headed straight for the shelter.

The gorillas, still wide-eyed from Jumoke's display, stood dumbfounded momentarily before they scattered wildly, as she blasted into their midst. Zuri tossed his hay-towel aside, and others who had begun to pick up prized bits of food when

Jumoke crashed onto the scene, abandoned all in their haste to clear out.

Jumoke ran a semicircular course that traced the interior shape of the shelter, swinging the top of the tree behind her and leaving a shower of raindrops and leaves flying through the air. Going full tilt, she headed back out into the rain. She heaved the treetop into the air, and the other gorillas, who had hurried back into the shelter, looked around wildly, trying to guess what she would do and where she would run next.

As they stood with their mouths open, Jumoke leaped into the shallow stream running through the habitat and sent water splashing in all directions. She straddled a log that joined the two sides of the stream and splashed water onto it. After doing this several times, she beat the log with both hands while making strange vocalizations.

Then inexplicably she went limp, lying face-down on the log, her arms and legs dangling on either side. Again the gorillas and I watched intently for signs of life. After many moments she casually dismounted the log, careful to avoid the stream, and sauntered back to the shelter as if nothing had happened. I thought I had never seen such a cleansing ritual.

Jane Goodall, having watched similar "rain dances" among chimpanzees, speculated that such displays are very old and have their beginnings in an ancestor the apes share with humans. She believes that these displays gave rise to the elaborate rituals and ceremonies we still associate with the changing of seasons and the marking of time, of the patterns in our lives. *So this was religion,* I thought.

In my explorations on the topic of religion, I knew that

theologian Joseph Campbell believed in an early genesis for ritual, perhaps even earlier than the ape-human split; and that ritual involving elaborate motion patterns is a very ancient behavior, perhaps marking not only tangible events but rites of passage as well.

As Jumoke entered the shelter and approached Nina, she rested her head on Nina's shoulder and put her hand on her protruding belly, ripe with life. Jumoke's ritual seemed to indicate that she realized that a time in her life was ending and another was about to begin as a new member joined their number. This was something I understood. My years with the gorillas were one rite of passage after another.

I had never been a religious person. I felt uncomfortable in most churches, finding no comfort in those buildings. Perhaps, I thought after watching these gorillas and coming to love them and learn from them, I needed to expand into God rather than being enclosed by a church. On reflecting upon gorilla ritual, I realized that my conception of God—as a great spirit living in things that could absorb me as I flung myself outward—was probably much the same as a gorilla's experience of God.

A God that can contain you as you expand into tree, sky, and rain can certainly also draw your being into a pinpoint of concentration. The expansive and the tiny—we are made in its image. It is our nature.

I was familiar with the spirit that drew me in. I had long been criticized for my ability to close the world off and focus my mind like a laser. I learned to admire the gorillas' ability to focus with the same intensity at times, and therefore I began to feel positively about this ability in myself. There is a certain beauty in seeing, as the saying goes, the universe in a grain of sand. There is also beauty in seeing a tiny universe, ripe with possibilities, unfold before you.

This was my feeling when I had, after years of study, attained my M.A. and decided to celebrate. I was excited about the possibilities before me and happy to be where I was in life. I felt comfortable being alone and rarely felt lonely. I enjoyed my own company. I bought tickets for an evening cruise around Puget Sound and took myself out to dinner before I boarded the boat. I was happy keeping to myself; not only was it natural for me, but I wanted to think about my life, my future. I staked out a spot on the bow of the boat and tried to ignore everyone. I exchanged a few polite sentences with people who tried to start a conversation but I really wanted to be left alone. When the sun dipped low and most of the people went inside the cabin to drink and dance (*which one can do anytime on land, so what's the point of taking a cruise,* I thought), I was glad to finally be by myself. I breathed deep and stretched out as the boat slid through the waves toward the amber-colored end of the day. As I was thinking how wonderful it would be to be left alone forever, I saw someone standing still out of the corner of my eye. I wanted her to go away. *Don't make me talk to you,* I was thinking.

But the person just stood, quietly, looking out at the water. I risked a more direct look. She seemed cold as she hugged her arms to her body. I turned back around and pressed my lips together. *I'm not talking to her,* I thought. After about ten minutes I turned around again. Even before I looked, I knew she was still there. *Oh, for Pete's sake.*

"Do you have a coat?" I asked.

"I have one inside," she said, and wrapped her arms around herself tighter.

"Do you want mine? You can use it. I'm not cold."

So, stiltedly, we began a three-hour conversation that was to prove the best I have ever had in my life. I talked about all of my ideas and poured forth my passion and listened to her ideas with real interest. I was sad to get off the boat and say good-bye. She said she would call me, but I had learned that that was something normal people say and it is understood that it doesn't have any real meaning. Somehow, though, I knew she would. She did.

She was very shy about asking me on a date but finally managed to get the question out. I quickly accepted. I was excited about the prospect of spending time with someone who enjoyed thinking as much as I did. It had never occurred to me that I could date women that I actually had something in common with. However, I was very cautious. I enjoyed my life and my career and I didn't want to give that up to please someone else. I had not had a pleasant emotional history with relationships, and I didn't care to repeat past mistakes. Though I knew romance was important, I also had come to believe one should approach such situations logically.

Armed with a desire to balance both aspects of my new foray into dating, I inadvertently provided Tara with the oddest date she had ever experienced. Things started out predictably enough. I took her for a ride on my motorcycle so we could see the setting sun over the city. I had packed a romantic picnic with sweets and strawberries and took her to one of my favorite places—St. Mark's Cathedral—and we sat on the hill behind the building to look out over the twinkling lights. We eventually went to a movie. As we sat in the theater waiting for the show, I decided that I had demonstrated my capacity for romantic overture and it was now time to get down to brass tacks. It was time for Tara to pass "the test" before I wasted any more time pursuing a doomed relationship.

I proceeded to grill her about her views on a host of issues: What do you really care about in life? What are your goals? How do you feel about animals, the earth, our place in things? What have you learned from your past relationships? How do you relate to your ex-partners?

Tara took this sudden change in demeanor in stride (though later she admitted it was disconcerting, referring to our first date as "the inquisition"). She told me she cared about being a good person and a good mother to her son from her previous marriage, that she enjoyed teaching English at the college where she worked but was also open to exploring new ways of living. She told me she felt that deepening her spirituality was very important to her and that she saw animals and the earth as an important part of that journey. She was still very close to her ex-husband, and their divorce had been very respectful. She said that she believed it was important to honor people she had been with even though the relationship might have changed.

It's interesting, but sitting there, listening to her answers, I suddenly realized she was beautiful. Maybe because it is hard for me to process perceptions when I am under stress, I have a difficult time with faces. I think that as I relaxed with each thoughtful answer she gave, I was able to take in more stimuli. I was taken aback by the fact that anyone would have said she was physically stunning. I remember thinking she possessed an embarrassment of riches: she was smart, funny, thoughtful, responsible, sensitive, and gorgeous. The bonus was that she seemed to understand me intuitively.

She laughed at my jokes, listened to me, told me her secrets, and cherished my difference. She did something else, too. As we continued dating, growing closer and closer, she started interpreting human behavior for me. If we would go to a party or overhear a stranger's conversation, I would ask her why something was said or why something was done, and she would tell me. Just like that. If I made a social mistake, she would tell me and explain what I did that was out of place and why it was so. Although for several years after I met her I was still cornering people with my uninvited presentations, I slowly started to memorize rules and even understand their structure and purpose.

I have read about the relationship between Helen Keller and Ann Sullivan. Sometimes I think of my relationship with Tara and believe it is similar. Tara insists that I give her just as much, reveal just as much to her. Perhaps this is what real love is—each person feeling that they receive the gifts of a relationship while giving such gifts without even knowing how. Perhaps in defying normal boundaries, love can create us in its image.

CHAPTER 7

Requiem

There was one more lesson I would learn from the gorillas.

One morning I came into work at the zoo and could tell immediately that something was very wrong. The area smelled different, and a strange silence hung in the air like the heavy, charged grayness between hope and resignation. I walked into the indoor cages and saw Congo, shrunken into himself, immobile on his hands and knees amid the soft hay. He cradled his huge head in his hands, looking intently at the floor. He was having a conversation with his death.

He was clearly in pain, and no one was sure what was wrong with him, though the vet had been up twice to try to help. I was left alone with him. It was like living in a snapshot that will continue to exist after everyone has died. My senses were razor keen: I smelled the sweet hay, and all the colors stood out brightly . . . it was humid, and there were strawberries in the sink. I could hear Congo's breathing. I knelt down close to him and sang "You Are My Sunshine," the song I always sang to him, which he had always grunted along with, ending softly: *"Please don't take my Congo away."*

He managed to grunt once because he loved me. The way he loved everyone. In reaching out so openly, so trustingly to all who came seeking a chance at love, he was everything—through a natural outpouring of heart—that I will never learn to be in a lifetime. I asked him please not to die. I cried, sitting next to Congo with tears dripping onto my boot.

Finally, I had to leave and go home. The next day I was listening to an old hymn when the phone rang. I looked out on the morning, fresh from its emergence from the night. I heard the song both in my ears and in my soul: *How can I keep from singing?* The phone rang again. I didn't want to answer it, but I did anyway, because nothing could stop it. When I heard the voice of a co-worker at the zoo, I wanted to hang up to make what she was going to say not true. But I stood still, and she said it.

"He's gone."

Her sobs receded as she said she was sorry and then hung up. I stood there frozen. Some part of me believed that if I didn't move, I could just hold my place, like a bookmark, so that someone could open time and move back to the pages he had missed, perhaps to put the book down again and forget to go on to the part where the world shattered.

When I went to the zoo later, bouquets were already laid against the public viewing windows, with cards saying "Sorry . . ." It was as deserted as a hole in the earth freshly dug in the middle of night. The funeral had no mourners. Somehow it seemed fitting that no one was there—the absence mirrored his great spirit, which had left evidence that it had lived. But then it disappeared to some other place where people don't stay to stare.

When I went into the back, nothing had been moved much from yesterday morning when Congo had suddenly collapsed and gone into a seizure, bringing vets and technicians to swarm and clamor around his thinning spirit. I looked at the spot for a long time. It was just a spot, a small place. No place on the earth would ever be the same.

The other gorillas cried for over a week, and their wailing filled the halls and spilled over the glass. They were lost. I felt lost, too. Through the gorillas I had learned how to feel and share so many emotions; now I knew what grief was.

Human people die, and those who survive them are sad, but human people have choices in life. A death such as Congo's—that of a deeply forgiving person who knew no choice—demands devastation. His death, overwhelming in its own right, represented to me all the pain and loss that I understood and the grace that I found through his forgiveness and the acceptance of the Gorilla Nation. In this way the devastation—though present as itself still from time to time—has been transformed for the most part into a certain kind of joy.

Bringing Congo Home

After Congo's death, I lost the heart to go back to the zoo. Before I left, I spent a last day with the gorillas. I helped feed them and spent extra time being near them in the office. I felt like crying. It felt good to want to cry—they deserved my tears. I closed my eyes and felt a line of wetness squeezing out. I smelled the gorillas and tasted the hot, grassy air hanging humid in my mouth. I listened to their shuffle through the hay and the clang of the doors that released them from their night rooms—and from me. I felt. I felt.

When they had gone out into their habitat, I walked outside and down the path. I stayed and watched them, alone, thinking about all they had taught me. I thought about the rebirth I had experienced in their midst. *Now I can remember two of my births,* I thought. I remembered a poem I had written just before I met them.

Our dust swirls in the slow brush of the passing years,
And settles in our dreams when the breeze pauses here,
Rising in the current of new life,
Making clouds behind your eyes,

And you came to me one evening as I was standing in the half-night,
With one hand you held out light,
And with the other hand you took my blood and roses,
A gift I gave to see your red dreams,

That is how we stood at the end of the day,
In the crimson of the sunset,
The crimson of our chase,
The crimson of our lives,
Leaving and returning like the sun.

I thought about Africa, about how Africa should have Congo back, have his raw material, have his bones to build on, perhaps to whisper his spirit home to stay, to rely on his heart to save the Gorilla Nation. Now that I was leaving, with the confidence and relative peace I had gained from the gorillas, I knew I could follow any dream I had. My first dream was to take Congo home. Traveling to a place far away and very different is daunting even for people without a dis/ability, but for me I knew it would be nothing short of a rite of passage. I was going alone in my body, but I knew I would be surrounded by the Gorilla Nation: I was going to the home that our ancestors shared, and bringing Congo home, and that was what I focused on as the date drew near. For me it was a symbolic journey toward a wholeness I had not known before.

But there were complications. When I asked for Congo's ashes, I was told that another zoo still owned him. They planned to take him apart and send pieces of him around the country for analysis. They still *owned* him, you see. I could tell by people's voices and faces that this was supposed to be a really simple thing to grasp. I still can't think about this even now without breaking into pieces along with him.

One of the people who had been there when he died opened my fingers and put in a little ball of hair. She told me that the vets had shaved it off trying to get an IV into his arm as he died, and she had saved the hair from the place they shaved. She told me to take this tiny part of him home. And so I flew to a place that was as far away as was possible without leaving the planet. It was as far into the human world as I had ever dared to venture. I would have done anything for Congo.

Once I finally arrived in Africa I found a beautiful place

under an acacia tree full of weaver nests. The sun was bright, and a heat like Congo's living breath surrounded me. I sat down and found a stick, and I dug and scraped at the powdery red earth until I had made a hole. I thought about how interesting it is that we come out of a hole and then go back into one in the end. I reflected on the fact that the hole I made was about the same size and depth as the one he came to us through. I put his hair into the bottom and looked at it for a while. It seemed fragile. I covered carefully. I found a piece of bark near where I sat. I wrote "Congo" on it and put it on the fresh mound of earth. I sang a song for him and then got up and walked into the future.

So that other people could remember these moments at the end of my friend's life, I wrote in my journal that night:

> I sat there staring at the only part of him that will ever touch Africa again—the first part of him to touch it in thirty-eight years. I pictured his wide, innocent eyes blinking in wonder at the bright African sun. I said, "Welcome home." And then I covered up the hole and planted the little plants back so it looked just like it did when I started.

And so it is when we pass our thresholds, and the persistent spirit of a healing world covers our long tracks.

Part Three

HOW CAN I KEEP *from* SINGING?

CHAPTER 8

Ancestral Hymns: A Family History

To you, the reader, it must seem odd that I could go through all these changes and experience real growth without ever knowing that I had Asperger's Syndrome. Obviously I knew I was different, but questioning how an autistic person could go undiagnosed until adulthood seems perfectly reasonable. This is actually a fairly complex question, with a complicated answer. As I mentioned earlier in this book, Asperger's Syndrome didn't make it into the *Diagnostic and Statistical Manual* (*DSM-IV*) until 1994, and as I also mentioned, I developed a lifetime pattern of using my intelligence to find ways to appear normal—to "pass."

Had things been different, any number of my family members might have been diagnosed with specific disorders. Most of my relatives, including my grandparents, aunts and uncles, and mother, all have severe allergies and serious respiratory illnesses that themselves have been linked to autism spectrum dis/orders. Though professionals have documented this pattern, they have yet to speculate on why this phenomenon exists. My personal belief is that the link is elementary: people without filters have problems with their environments; whether it is light or sound or pollen or dust, their bodies react strongly in defense. This is what I think leads to many diseases, in this case, autism, asthma, and allergies. Beyond the physical problems, my family's mental history has, if current diagnostic procedures for autism are taken into account, much relevance.

For example, my maternal grandmother was agoraphobic and barely left the house for many years before she died. She engaged in terrific monologues that would continue even after you had gone out and let the screen door whine and slam behind you. She had all manner of nervous tics. She was a wizard with numbers. She had chronic insomnia. Her sister, my maternal great aunt, was nicknamed Buggy because everyone thought she was a little crazy.

My grandfather was obtuse. He once dropped a biscuit on the floor and, after standing there for a moment, dumped the whole pan of biscuits onto the floor "to make it worthwhile to get down there and pick them up." Though he was as gentle as a lamb with people, he attacked and demolished a windup swing after trying patiently to read the assembly directions but failing to put it together. He threw a toaster out the kitchen window one morning when it wouldn't work. His brother, my great uncle Charlie, was said to be "retarded" and had bouts of temper also. He didn't speak much at all, and it is unclear whether this was due to choice, capacity, or mechanics, as he had fallen with a stick in his mouth and injured his larynx when he was a young child.

My maternal uncle had a plethora of unusual skills. He could draw a three-inch line—without the use of a ruler—to within a tenth of an inch every single time. He could carry pi to two hundred places, writing the numbers down straight from his head onto a piece of paper. He could balance things—one of his favorite pastimes. Whether standing at the kitchen counter or sitting at the table, saucer would follow cup, topped one after another with pencils, salt and pepper shakers, spoons, knives, and glasses. My grandmother would yell at him that he was going to drop and break everything, but his concentration was unshakable. He had a uniform: jeans, denim shirt, and black polyester sweater vest, which he wore every day I can remember during my childhood. He had a tic that consisted of a grimace and a shrug of his right shoulder. I always liked it and never found it odd.

My mother is very methodical and has always gone about her daily routines with a generous portion of ritual. She has always liked things done a certain way and in a certain order. She has suffered from depression and sleep disorders off and on in her life. She has an absolute loathing of change and must be prepared in advance for any activity and certainly any revision in plans. She likes best to be alone and lives, with my father, on twenty acres in the deep woods of Montana, where she stays most of the time, preferring her daily communion with deer, rabbits, birds, and squirrels to any human company.

My father, prone to attacks of rage as a younger man, has mellowed and gained peace through his mostly solitary life with my mother on the mountain. He needs order and has very specific places for every piece in his many collections, which range from Marilyn Monroe to tin toys, first edition books to antique science paraphernalia. He has always been upset by chaos, by slow and uncritical thinkers, and by too much company. Once when I was visiting, I was swinging in the porch swing, and he was standing beside me. He clutched a pair of binoculars in his hands and pressed them fretfully to his face. His brows were knitted, and he said, his voice tight with tension, "See! See down there? Neighbors!"

My father's father was a lifelong substance abuser, as were my father's brothers, both now dead from hepatitis C. His mother suffered from panic attacks and never made friends easily. She also preferred being alone to being in company most of the time.

Finding a Name for It

The period when I grew up was a different time, though, and my odd family members were more often accepted and absorbed into society. Their quirks were even appreciated, in ways they might not be today, especially in an already eccentric family. Additionally, I come from a working-class family descended from coal miners, and working-class people more often than not don't have the money to spend on expensive diagnoses, which they often see as having little practical value. Among the working class there has historically been, and remains, a deep distrust of doctors in the field of psychology. Add to this mix my parents' reluctance to deal with anyone, and it is no surprise that my behavior was simply tolerated.

It wasn't until I was grown and had had my own experiences with college, different class backgrounds, and different ways of thinking that it occurred to me that that which had always made me different might have a name. As with many adults who are eventually designated as having autism, it took a final, additional step—having a younger family member diagnosed—before I sought an official identification. This is in fact a common pattern. Autistic traits tend to run in families and then distill into individuals, who in spite of certain impairments produce offspring who also have a full-blown autism spectrum disorder. When these children are identified as autistic, the pieces of the puzzle suddenly fall into place for others in the family.

In my situation, a younger relative was long faced with profound difficulties similar to my own. He grimaced and held his body in strange postures, spoke to people in ways that were blunt and considered rude, defied authority in any form (and he was exceedingly bright and capable of great logic in these situations), and displayed nervous tics when his many anxieties and sensitivities were brought close to the surface. Like me, he did poorly in school because of these problems and the challenges of impulse control. Unlike me, he was aggressive with people—people whom he told me he experienced as blurry objects exploding with invasive stimuli—and though I understood better than anyone why he was doing these things, a continuous battle ran in my mind. Watching him, I gained insight into my own behavior and motivations, yet seeing him act in these ways was painful, like looking in a mirror. As I watched him, I was often filled with rage. I was embarrassed because it reminded me of my own behavior and

I saw myself the way people must have seen me before. I was desperately sad thinking about what I had gone through internally, feeling my isolation all over again. I swung between embracing my young relative wholly and rejecting him wholly. I felt compelled to help him but just as repelled by being near him. Unfortunately we had, many times, entrenched opposing needs: I needed absolute quiet, but he would cough loudly and compulsively; I wanted to be invisible and avoid the piercing stare of strangers, but he always brought attention to us whenever we were out in public; I needed absolute order, but he found comfort in fondling objects, only to let them drop wherever he finished with them.

Finally, I felt I could only help him from a distance. Armed with a deep identification with his subjective experience, I began to do research in order to put a name to his challenges. I matched his behavior to a narrowing field of possibilities, all the while comparing my own experiences with his. I admitted to no one, not even myself, that I knew what his problems were because I had them myself. When Asperger's Syndrome was arrived at as the cause of his behavior, I knew it was the right diagnosis. A professional diagnostician soon confirmed that this was indeed what was going on with him. I remember her saying that other people close to him might well have the same thing, because there's a strong genetic component, but at the time I pushed that possibility aside, even though I knew that I was the most likely of all to qualify. Whenever my mind wandered in that direction, I kept coming back to certain things I had been told: only boys have Asperger's, people with Asperger's don't have real emotions, they don't have friends, they don't have partners in life, they can't relate to other people at all. And then there was the aggression. That was something I absolutely couldn't relate to.

Accepting Asperger's

Ironically, my relative's diagnosis made me even angrier, probably because he began to get services at school, increased understanding at home, and a framework for explaining himself *to* himself and to others. His intelligence was focused on as a great strength and the thing that could save him. He was loved and understood. Although I experienced a vicarious joy at his new window of knowledge and understanding, another part of me was angry that I had not had the same window.

My agitation grew and boiled over. Whenever I had occasion to be around my young relative, I would eventually blow up. It was just a matter of time—sometimes in the midst of the family visit and sometimes a week later, but it always came: I would feel a sudden rush of adrenaline, as if I were being scared to death by thin air; then have the urge to shove everyone and everything away from my body; then feel the crushing weight of being buried by everything. I would feel sad and ashamed, even as I stomped around screaming at the top of my lungs. I had broken my own moral code and was worthless. I couldn't control myself and deserved to be scorned—or at least that is how I felt behind my glaring eyes, shouting mouth, and balled-up fists. It was like being in a car that had lost its brakes and looking on in horror as I mowed down pedestrian after pedestrian. I began to alienate everyone. I felt at times that I had lost all of the peace that the gorillas had given me.

My partner, Tara, bore the brunt of these rage attacks. A kind and patient person by nature, she would try to remain calm and supportive through the inner storms that I would suddenly turn with full force against her. I couldn't bear the criticism that the truth leveled at me, and so I blamed her—or the world, or the weather. It was always the same. I would almost always come to her later and offer a mechanical apology. I felt hollow when I did.

One day, as I screamed and fumed and slammed the door, a look came over her face that I had never seen. Very quietly she opened the door and said, "I don't want to be with you any more." I knew she meant it. I knew in a moment that I had ruined the safety, love, and stability that I thought I had won. I was about to lose everything I had gained.

I asked her not to go. I told her I would find out what was wrong with me. I told her what I should have admitted all along: that I couldn't cope with my feelings, that I was terrified and overwhelmed all the time, that I didn't think like other people—in short, that I deserved criticism but hoped for understanding. She agreed to stay with me if I got help. I set about the task of getting a diagnosis. Since I knew what it was going to be and was finally facing my fears about it, it was a fairly easy matter to get one. I meticulously documented every autistic feature that applied to me from birth to the present. I called my parents and asked them questions for two hours, writing down all that they said on a form that I had constructed. I believe that they were relieved to know that there was an explanation for all that they had gone through; at the same time I was moved by my mother's frequent assertions during the interview that, though I was odd and difficult, I was special, and she and my father had always loved me. I told them not to apologize for telling me what they had seen and thought; it was going to help all of us now.

After I interviewed my parents and sister, I tabulated all of my own detailed memories. I created complex charts and accompanying narratives demonstrating that I fulfilled all the criteria for an autism spectrum disorder. I even found and edited audiotapes made when I was a child to demonstrate my enormous vocabulary, odd prosody, and flat affect. Then I called a woman whom I knew to be qualified in the area of autism spectrum disorders.

"I need a diagnosis of Asperger's Syndrome," I said. There was a pause, then I continued. "I know you have expertise in this area, so I would like to see you." There was another pause.

"All right," she said. "What does next week look like for you?"

When I went to her office, I brought along my huge folder of documentation, and before she could say much of anything, I went through and explained each chart and entry, and the notation system I had come up with to make cross-referencing easier. I handed her a separate sheet detailing the reasons I needed a diagnosis. Throughout my presentation she wore an expression I couldn't really read. Then she told me we would have to have several visits to determine if I did indeed have Asperger's. I was somewhat disappointed that I wouldn't leave with a diagnosis, as I felt I had made her job simple. During our last visit she thoroughly went over her findings, then laughed and said that the way I had presented the material to her was almost strong enough in itself to ensure a diagnosis. I was puzzled by this. I still think it is silly for people not to be prepared when they work with a mental health professional.

Though gaining the diagnosis took longer than I had wanted, the end result was the same. Sitting in my car in the parking lot after the last session, I felt an immense wave of relief wash over me as everything suddenly made sense. I looked back over my life, perhaps the way people do before they die, and thought of all the painful memories that could now be explained. Like someone making amends in a twelve-step program, I almost felt compelled to contact everyone who had ever been impacted by my autism—whether positively or negatively—and explain. It made me feel both better and worse knowing that I hadn't meant to disturb or hurt anyone.

When I meet the family members of someone with autism who haven't told their loved one—usually a child—of the diagnosis, it perturbs me, because they want to "avoid labels." I can assure you that not only does the autistic person always know that they are different, but they suffer deeply from not knowing why. While they try to come to understand themselves without having a name for their condition, other people definitely *are* labeling them—and usually without the compassion that real education would bring. I have been told that the people in question want to save the autistic person from the "stigma" of being autistic. I don't think they realize how stigmatized people with autism are—because of their behavior, not because of any label—and if they don't know what is going on with themselves, their behavior isn't likely to change. Further, autistic people are, as a general rule, not likely to care what other people think. Perhaps this is because it is so hard to figure out.

I took the several-page report home and showed it to Tara. She wasn't at all surprised, but her initial reaction was not positive. The thought that I was "officially" impaired did not appeal to her. For her, it was almost as if the diagnosis made it inescapable that I would never be able to connect with her deeply, to really be there for her or be able to change. It is hard to blame her when the cut-and-dried description provided by the *DSM-IV* makes people with Asperger's seem to be the epitome of cold disinterest, complete uncaring, and total self-absorption. What Tara learned, and what I hasten to remind the reader, is that the characteristics described in the *DSM-IV* are just that: they are descriptions of *coping behaviors* and not descriptions, necessarily, of innate orientation. People with Asperger's seem not to want to reach out, but it is not always a problem of desire, but one of comfort: they need to feel at ease in their bodies and at ease with people they might be interested in knowing, for instance. In the past I have felt unable to reach out to people because I was physically uncomfortable.

Wanting to demonstrate to myself as much as to Tara that I didn't have to live as a diagnostic description, I laid out my plan of action. I researched medications for the anxiety and obsessive/compulsive components of Asperger's. I had heard good arguments against medication, especially where growing children are concerned; however, I believe that the long-term damage that can result from adrenal stress may be worse than side effects caused by medication. I found over time that the best combination of medications for me was gabapentin (an anti-epileptic first designed for children, with a low incidence of side effects), bupropion (an antidepressant that targets my obsessive/compulsive tendencies), and alprazolam (a tranquilizer that helps me sleep by diminishing nightly panic attacks and my need to move my feet and legs constantly). I began a journal to help me reflect on my actions and piece together a deeper understanding of social cause and effect. I made a list of comfort activities (like going out for root beer or getting a head massage) that I could engage in when I felt a rage attack coming on. I also allowed myself to use the comfort of some of my more innocuous rituals to help me keep balanced so I didn't go through the typical extremes of "appearing normal" and then falling into incapacitation.

Another strategy I had was to join an online discussion group for adults on the autism spectrum who have been to university. Many of the participants shared their difficulties with such honesty and trust. Our discussions meandered from the difficulties of learning in a traditional environment to the problems with showing love, from what we hate most about surface social niceties to how to change the world to make it a better place for all living things. Through insights gained in the group I was able to discern what parts of me are likely to be driven by autistic factors in my makeup and what is just quirky because I am a unique individual. Equipped with this awareness, I started to tell other people about how I thought and felt about things and explain why I did the things I did.

The inspirations of those revelations led me to work to

bring books on these subjects into public consciousness. In my desire to help the world improve, I am not unique as a person with Asperger's. But the gorillas have given me strength to see that if I am to be effective in bettering the planet, my responsibility outweighs my comfort. I think this is true of every human person.

Sometimes I feel like I am now the one in the zoo. Like any good gorilla chained by what is often unpleasant circumstance, I have evolved to incorporate endurance as a keystone. People come to hear me talk, to see how I behave, to see what I look like up close. Most of them give something back to me and make it easier for me to continue to reach out, to communicate in small ways (like taking the time to talk to my grocery checker), or in large ways (like writing books that touch thousands of lives).

CHAPTER 9

Successes

Sometimes reaching out and communicating isn't easy—it can bring sadness and regret. Some of my family and friends, after reading the manuscript for this book, were deeply saddened to learn how I experienced my world. They pointed out that I had had some happy times, despite the sadness in the story. This is true, and my recollections of those times are as vivid as my memories of sadness and isolation. This story isn't always about the times I was happy—it was written to document the journey of an autistic person from birth to adulthood, and the hard memories, unfortunately, are the ones that seem pertinent. However, I have had happy times.

I have heard other autistic adults say that, despite islands of happiness in their childhood, isolation, confusion, and depression underlay even the best memories; it wasn't until adulthood that they realized there were many good things about their uniqueness. I agree: I enjoy my life more and more as I mature. I think it is important to share these times with people and give them hope.

Finally: A Wonderful School Experience

During the early years of our relationship Tara and I both applied to and entered graduate school. I searched the world for a doctoral program that would permit me to work in the solitary manner I found necessary and to honor the connection with the gorillas that had brought me so far. At that time Europe was far ahead of the United States in having progressive ideas and interdisciplinary degree programs. Additionally, that area of the world has a long-standing tradition of mentoring approaches over coursework as we think of it here.

My research with gorillas and a developing fascination with the "core ideas" or archetypes that shaped both our emergent cultures led me to read a great deal of the work of Carl Jung, which in turn led me to explore universities in Switzerland. After much searching I found a university that fit my criteria. But I was told that because their standards were more stringent, they would not fully accept my previous degrees; I would have to test into the university and rewrite my M.A. thesis before beginning their doctoral program in interdisciplinary anthropology.

I overcame my worst fears about going to a foreign country and being without the comfort of anything familiar. I braced myself and made the trip to Switzerland, and then to the university, where I completed my oral exams, documented and explained my undergraduate studies, and exhaustively related the methods and findings of my previous research. I defended my new thesis and detailed its relevance to my proposed doctoral work. When this was finished, I defended my dissertation proposal. I passed my exams, and my doctoral program and dissertation proposal were approved.

When I returned to the States, I chose my advisers, and after they were approved, I worked on my doctorate for the next several years. Proudly, I completed my defense with the university's highest marks. I felt that this triumph was as much the gorillas' as mine.

Once I found a university that supported my learning needs and the direction I wanted to go in regarding my dissertation, graduate school was a fairly easy experience for me. But this wasn't the case for Tara. She had been involved in a graduate program before, and it was full of bad memories. She had been married to her first partner (a man who was working toward a Ph.D. in philosophy), and they both attended the University of Rochester. Though the environment was one her husband excelled in, she had felt alienated and lonely. He had been reluctant to start a family, wanting to finish his program and find a job first, but Tara very much wanted the intimacy of family connection, and they agreed to try for a baby while they were still in school. Unfortunately, the isolation, depression, and hormones all combined to make it virtually impossible for Tara to continue her own studies. The worst of it came when Tara undertook her oral exams and failed. She knew the material, and her committee knew she knew it, but despite their supportive coaxing, all she could do was stare at the table, lost in a blank and unable to speak. After trying many times to get the discussion started, they reluctantly informed Tara that the exam was over and that she hadn't passed.

To her great credit, after getting a divorce and finding herself years later, she decided to try again. It is a rare and difficult thing for a student to come back successfully after such a setback, and Tara knew it wouldn't be easy, but the people at the University of Rochester had liked and respected Tara, so when she reapplied to graduate school they were happy to take her back. Tara was once again a student. I admired her resolve, and when she said, "I'm sure there will be times when I will want to quit. I'm going to count on you to help me keep going. Just remind me of all the reasons I *need* to do it," I readily agreed.

Only a few weeks later she was feeling overwhelmed and went for a walk around the lake near our house. She came back smiling. As she strode into the cabin, she was clearly at peace. "Well," she announced, "I realize I don't have to do it. I can just quit. I don't have to put myself through this torture."

I was in a real bind. She seemed so relaxed and happy—yet I had promised to help keep her going. I started to list all the reasons she shouldn't quit. As I forged on, she started to cry. I held her as she sobbed. I felt terrible and wondered if I was doing the right thing. But she agreed, and it was the last time she seriously considered ditching the program. Now she is glad that I pushed her to press on. It didn't take long for us to be glad of the decision, as it hailed one of the best times in our lives.

We had long talks into the night about our ideas and the

arguments we were working on. Our areas of interest complemented each other: at a time when most of academia was rushing to embrace postmodernism, Tara and I both held an interest in essentialism. Tara was writing her dissertation about Native American gender roles as they were explored in Native American writing. She illuminated the phenomenon of seemingly innate identities in certain indigenous cultures and the ways that they were recognized even in the very young, who were seen as falling into certain gender categories and were trained for their cultural roles accordingly. I found it fascinating that alternatively gendered youth were thought of as special and were instructed very early in the ways of their predecessors and in the responsibilities of acting as doorways into the spirit world. Tara believed there was strong evidence that certain people were born with this special kind of ability and knowledge, giving them an essential platform to build on as they studied formally under their elders.

For my part, I was looking at the possibility of ancestral memory as a physical phenomenon that existed as an environmental component rather than as a discrete brain trace—in other words, the possibility that our minds and memories are not specific to us as individuals but are part of a resonant fabric that informs our psychology and morphology.

I included theoretical physics, anthropology, Jungian psychology, philosophy, and folklore in order to bring a strongly interdisciplinary approach to looking at the patterns inherent in stories of the beginnings of humankind. My secondary focus was the universal story of the archaic "ape-man," who, according to so many cultures, was the first person.

Tara and I would work eighteen hours a day on our

studies. It was a joyful time. We would sit near each other silently, reading, thinking. Then when it got very late, we would go into town to share a pitcher of beer, over which we would challenge each other, support each other, and feel the wonder as our ideas and approaches became more than the sum of their parts. It felt so good to have a perfect friend to be with and discuss things with, no matter how "strange" my ideas were. To have nothing to do but think deeply and with discipline—at least for several years—was wonderful.

Though she was extremely nervous. Tara boarded a plane to the University of Rochester for the defense of her dissertation. She did beautifully. She even enjoyed the defense, as it gave her an opportunity to share her thoughts with people she had long admired and enjoyed. She had come into her own.

My own defense was a strange affair, like something out of a David Lynch movie. I had assumed I would go back to Switzerland for my defense, but it happened that my committee was going to be in Las Vegas for a conference. Though it was somewhat early in my process, I completed my dissertation in anticipation of their invitation to defend in the United States. I booked a ticket to Las Vegas. There was a surrealism present from the beginning as I got off the plane to the sights and sounds of slot machines flashing and blaring in the airport. Tired-looking women with bleached hair hung limply from their stools praying to their stingy gods and asking for deliverance in the form of the shiny quarters that were only rarely vomited up from the bowels of the unblinking deities. Couples who didn't seem like they were in love looked at everything but each other, shell-shocked expressions on their faces. I took a cab to the hotel where I was to give my defense and tried to go over my notes once more amid the brightly colored posters of kicking showgirls with impossibly wide smiles, the clank and whir of the stealing machinery being plugged with even more small round pieces of metal, and the garish decorating that passes for taste within Las Vegas city limits.

My defense was somewhat disappointing. Though there were several people there who I was glad to finally meet in person and get to know a little, when the defense actually started, I waited in vain for really tough questions. Tara, who had been immersed in the process of my dissertation writing every detailed step of the way, had given me a mock defense the day before, and it was very difficult. Somehow I wanted to have a measure of discomfort during the defense—one final push of my mind before I let it go and allowed it to become its own entity, floating in space. It wasn't that my committee wasn't smart—I later had a very challenging conversation about my favorite philosopher, Paul Feyerabend, with one of the committee members over a celebratory drink. And it wasn't that they didn't know the material—they had read and reread it. It was just that no one could possibly be as obsessed with the topics contained within the work in the way that I was. Perhaps this was a good first lesson as I left graduate school behind and walked toward my professional career.

Settling into Place

I was apprehensive about the next step of the process: looking for a job. Though I had originally gone into the program just for myself, my success both with people and as a scholar made me consider trying to teach. I sent out about thirteen applications and didn't receive a single call. The one place that had expressed an interest in hiring me was Fairhaven College of Western Washington University, but they received hundreds of applications for the job I had applied for, and the candidate they hired was not only beautifully qualified but an outstanding teacher as well. Ironically enough, this was a part of the equation I had never thought about carefully: the need to communicate well and have strong personal interactive skills with students. In retrospect, it would have been a disaster for me to teach a full courseload in a tenure-track job, trying to live through that enormous pressure while navigating the extreme social demands of a class and a department. Since I could barely handle being a student on campus, I don't know what made me think I could somehow make it teaching large groups.

Reflecting on the fact that the river doesn't curse the banks, I accepted that full-time teaching wasn't what I needed to be doing. When I asked myself what I wanted to do and what I would be *good* at, I sat down and began to write my first book in earnest. I had always promised the gorillas I would do something for them when I was finished with school, so I set about telling their story. Writing about the worth of their family and my time of connection with them, I started to embrace life on a new level. Since the beginning I had wanted the important parts of the life they had: connection, peace, context, family. Tara and I had talked about having a baby eventually, and both of us felt it was time. Though we didn't have full-time jobs yet, we had completed a list of goals we wanted to achieve before we had a family. We had finished our degrees and had a sense of direction. Though we weren't old, we were getting older. These were all good reasons logically, but I think we just intuited that the time was right. Whether it was the stars aligning correctly—some unseen cosmic settling in our favor—or just luck, we were right. Of all the potential people our son could have been, this one, the one we started then, was perfect. He has become my soul.

Starting a Family

When we decided to get pregnant, both Tara and I thought a lot about how we could undertake the project in such a way that would increase my influence over its development. We both knew that Tara should carry the baby: she has a perfect health record, had given birth with no complications once before, and actually enjoyed being pregnant. My history of asthma, my "general disposition" (as we called it then, before my official diagnosis), the fact that I had no desire to give birth, and the pharmacy that was in my bloodstream would not make for a hospitable environment for a fetus. I knew that I would love a child of ours whether I was biologically linked to it or not—I have always believed that genetics is highly overrated in the causality chain—and I knew that our child would not come to be if it weren't for my existence and that of my family, and that fact was just as powerful as DNA. Still, I thought a lot about my influence and found it exhilarating to consider how my choices would affect whom we brought into the world on a very fundamental level. For example, Tara and

I agreed that I should choose the sperm donor. I wanted him to represent me as much as possible, and I had a certain faith that I would find the right person.

Our infertility clinic used only one cryobank and I started looking for the right person as soon as we got the catalog of available donors. I selected several potential donors who shared my ethnic heritage and ordered "short profiles"—several-page questionnaires filled out by the donors—from the cryobank. The day I got the profiles in the mail, I tore open the envelope and pored over the selections. The first candidate, answering the question "What's your favorite animal?" had written "I don't particularly like animals." I tossed that one over my shoulder and out of the running. Though I didn't put too much stock in genetics, I couldn't take the chance that there was something pathologically mutated about this guy. The next candidate acted as though he were filling out a job application and was very stiff and formal, his words clearly constructed to impress. His major goal in life was to get rich quickly (and, it was intimated, at any cost). I knew he wasn't the match I was looking for. Neither were the candidates in the next couple of profiles. I was losing hope when I reached the profile at the bottom of the stack.

What struck me about it first was that it looked as if it had been written in my handwriting and in my style. The writer had ignored the constraints of the allotted space to answer and had written into the margins, even going sideways to finish his thoughts. Then I read his responses to the questions. He was intelligent, irreverent, open, honest, and self-reflective. His answers were also unusual. When asked what his favorite color was, he answered that he liked a particular subtle color on a particular kind of animal's back. When he was asked what he saw himself doing in ten years, he said he wanted to engage in space travel, raise an unusual meat animal, and run for political office. He picked a small and unobtrusive animal as his favorite, which I found touching for some reason. I knew I had found the right person.

I ordered the long profile and an audiotaped interview and through them learned a great deal about his family and his personal history. I learned more about his sense of humor as he related a funny story about a wild comedy of errors that he and his friends had played out one night, involving their car. Although I understood the rules regarding his anonymity, I wished I could know this person. He seemed like someone I would like to call a friend.

Tara and I went in for an ultrasound, and our infertility specialist determined which side Tara was ovulating from. We kept tabs on the egg's development over the next couple of days, and when it was nearly ready, Tara received an injection that would ensure that she ovulated within twenty-four hours. I had started a journal for our baby several years before we expected to try to get pregnant, and I continued it now, talking about the decisions we made and why, and my private thoughts about our child's conception and my own past. On the day of our appointment to get inseminated, I wrote:

> I got up at three this morning because I couldn't sleep. This is the day! I watched the sun come up while I thought about you. You will be the dawn of my new life. Your mommy and I went to the appointment at twelve, and I took pictures of the insemination. They put some

> sperm very close to the egg, so it should only take half an hour for the sperm and egg to meet! As I write this at 6:40, we might already have you growing. After the appointment we went to get veggie burgers and fries and onion rings at Barter's drive-in. We spent the day being close and hugging a lot and wishing you good luck in getting started.

Twenty-six days later I wrote:

> Your heart started beating today. We won't be able to hear it for another month or so, but we know from reading that *everyone's* heart starts beating on the *same day* after they start growing. If you feel your pulse as you read this, you can think about the exact day your heart started beating and remember it won't stop, and hasn't stopped since this very day as I write this.

That was the day our hearts started beating together.

Tara had a healthy and quiet pregnancy. At times she was difficult to live with as her hormones roamed the map. I finally gave up on trying to be rational and used a stock response that worked most of the time when she was at the end of her rope: "You're beautiful," I would say sincerely. "What can I do for you?"

Everyone was filled with anticipation when Tara's due date drew close. My sister Davina flew in for Christmas break on December 17. Once she was there, we were both so eager to meet the baby that we talked Tara into downing a dose of castor oil, which several people had sworn would induce labor if

Tara was ready. Trying to make it more palatable for her, I concocted a castor oil omelet (just chugging the stuff would probably have been more appealing), and Tara, who was also eager to meet our little one, dutifully complied and ate the whole greasy mess. Within hours she was having contractions. Not just in her uterus, though—she suddenly had the worst vomiting and diarrhea of her life. She announced from the toilet that she was going to die. I helped her into the whirlpool bath I had constructed from scratch, and that seemed to help. I rubbed her back ceaselessly to help her muscles relax. I gave her ice water and Popsicles. I turned on some soft classical music. I coached her softly, giving her words of encouragement. It was clear that this was the real thing and that the baby would be born in a matter of hours. We called the friends who we wanted to have there: one drove all the way to Tacoma from Salem, Oregon, and another drove from Portland.

By the time everyone got there, Tara was deep into labor and having a lot of pain. The worst part was that she was asking me to please help her, and there was nothing I could do. It did seem to help a little when I switched from soft reassurances to a more commanding platform, assuring her in a firm voice that she *was* going to live, that she *would* have the baby, and that everything *would* be all right. I managed to keep it out of my voice that I was trying to convince myself at the same time.

The midwife finally arrived, and I felt somewhat better but still had to leave once to go throw up outside. Seeing Tara in that kind of pain was so much worse than being in it myself. I remember whispering to myself on the way back inside that I would never, ever do this again. Maybe if *I* could have an epidural that worked from the neck up—that would be the only way.

> At 4:11 you were born. We all helped your mom get ready. She opened up from 3.5 centimeters to 10 centimeters in two hours. As soon as her cervix was open, she gave one mighty push, and your head came out. Another mighty push, and all of you came whooshing out. You squirted clear across the bed in a gush of water, some blood, and umbilical cord. Your midwife put you on your mom's tummy, and I rubbed you with a towel until you took several small, quiet breaths. We were all crying because we were so happy. We showed you the way into this world, and you will probably show us the way out of it. We all cuddled near you, and the midwife gave you a checkup to see if you were all right, and she found that you were perfect.

When my son was born, I sobbed. I had never cried so hard before. I was totally overwhelmed by the love I felt for him. I stroked his back and looked into his eyes as he pushed himself up on Tara's chest and looked around before settling into her and rooting around for a nipple. All I could say was, "He's so tiny . . . how am I going to take care of him?"

I'm sure most new parents have this same sentiment, and like other parents I found a way to do it and even surprised myself at how easily I was able to immerse myself into the routine of caring for my son. Tara continued to put in long days at work, now feeling pressure to secure a tenure-track job. It was hard on all of us for her to be gone so much. In response to the difficulty and unfamiliarity of my new circumstances, I fell back on my habit of structuring. I would wake up at six-thirty and sing to my son until he woke up. I would give him a bottle of breast milk and sing to him some more. Then we would get up, and he would play with materials that would stimulate all his senses. I would let him smell different spices and oils, and he always loved that, kicking his feet and giggling. I had a big sack of fabric scraps with different textures and patterns that he loved to grab and wave around. We listened to music. Toward the middle of the day we would crawl into the whirlpool bath together, and my little naked son would fall asleep on my chest for a couple of hours. While he slept I would write in my head, editing, adding, restructuring, so it would be ready for me to simply put on paper when I got time to sit down at the computer. When he was very small, he slept a lot, and I was able to write quite a bit. Sometimes I watched old television programs like *Dark Shadows* or *Lost in Space* and thought a lot about when I was little and my mother and grandmother watched those shows while I dozed or crawled nearby.

My son reached all his milestones early and seemed from the beginning to have a strong connection to animals. The first time he ever smiled was at our dogs. His first word was "fish" at eight months, a quiet "fssshhh" sound coming from his lips as he sat with me in the bathtub and held the porcupine fish he had always loved. He loved to be outside, to feel the trees and the grass. I knew I had given him the right name.

Teryk Brydanialun Prince-Hughes

Here's the story of your name. As long as I can remember, I have been interested in our ancestors and the people who came before us, even the ones who lived millions of years ago. I even talked to them inside my head when I was little, and they would talk back. That's why when I got older and went to school to become a doctor of philosophy, I decided to study our ancestors and the stories people tell about them all over the world. In Siberia they have a story about the Teryk, which means "Dawn Man." I liked the name because it has my name in it, it is one of the names for our ancestors, and it also means "One of the First People." Your name will stay with you your whole life, and it should say something about where you came from. That is how most people on earth do things.

Your second name is also very special. It is Brydanialun. It is made up of three Gaelic words:

bryd = heart
anial = wild
un = one

It means "heart of the wild one." One reason this is special is that it is the old language of your ancestors. My grampa, your great-grampa Eddings, used to call your gramma Joyce "Big 'un" and me "Little 'un." I know he would have a name like that for you too if he were still alive. The other reason this name is special is that when I was learning about our ancestors, I made friends with gorillas and watched and learned from them. We share the same ancestors, and by learning from them, I learned a lot about how our oldest ancestors lived. Gorillas have kept many of the old ways people used to live. Also, they are living and gentle. The gentlest and most loving gorilla I ever knew was named Congo. He and I were best friends. He had a very hard life but was kind to everyone. Especially me. I loved him in a way I will never love anyone else.

He died on February 27, 1996. I was very sad. The human people he lived with wanted to know what made him get sick and die, so they sent different parts of him to other people to look at and see what happened. I wanted to take him back to Africa where he had been born so he could return to the earth there. One of the hardest days for me was the day they sent his heart away for doctors to look at.

In this culture we always think of the heart as where people's love and feelings come from. I didn't like to think of Congo's heart going away. But after a while I understood that Congo's love was everywhere and lived in the people who love me. I love you in a special way I will never love anyone else, and in a way you are Congo's heart to me. You are the heart of the wild one.

I believe Congo's spirit lives in you and protects you. Maybe you will talk with him someday, as I used to talk to my ancestors.

Tara's Career

Though Tara had sent out around seventy applications during her pregnancy, she got only one call: for a temporary position at Pierce College in Tacoma. The thought of moving while Tara was pregnant was daunting, but we decided that it would be a good job to tide us over until she could go on the market again the following year. So we packed everything up, and Tara, going down on a weekend, found an inexpensive place that would take dogs. Unfortunately, it was in a pretty rough neighborhood. The houses, including ours, were drafty and poorly built. Our neighbors, who we never got to know, were always fistfighting the police in front of our house and regaling us with colorful expletives at all hours of the day and night.

Mostly I stayed inside. I rarely left the house during our entire year and a half in Tacoma. I did, however, finish writing two books and soon placed both of them with publishers. The isolation was difficult. There was a chance that Tara's temporary job could become permanent, and she often spent twelve-hour days at the college trying to become indispensable.

Though Tara did indeed have an opportunity to apply for a permanent job at Pierce College, a tenure-track position was being advertised back at Whatcom Community College in Bellingham, where she had taught for many years. There were several hundred applicants and we tried not to get our hopes up, but her interview went well, and it was clear that all of her old colleagues missed her and knew that she was an outstanding teacher. We were delighted to hear that there were three finalists chosen and Tara was among them. After an agonizing waiting period, we learned that the college was going to hire all three of the finalists for tenure-track positions. This was wonderful news, not just because we wanted the job so badly but because the other two candidates were already working at the college part-time and Tara knew them from before, so it was a great triumph for everyone to share. Our friends were so happy and put a lot of energy into facilitating our move home. One friend said she knew of the perfect house for sale just down the street from where she lived and took us to see it. It had a three-lot yard with fourteen big trees, including three apple trees. The entire backyard was privacy-fenced, and it was as big as a field and just as overgrown. The house had been built in the late 1920s and was small but had three bedrooms. We couldn't look inside properly, but we fell in love with it from what we saw. I had wanted to live in the country, and Tara had wanted to live in the city: this was the perfect compromise. When we met with the realtor, I was glad that he stood back and didn't hover as we looked around inside. I liked the house, but when I went to the kitchen, I felt spellbound. It was just like my grandparents'. There were three rounded shelves on each side of the window above the sink, trimmed with strips of silver metal. There was a wooden valance above and where the cabinet doors were; there were three ventilation slits just below the sink. It was like walking up to the window in my grandparents' kitchen. I wanted the house, and Tara was equally sold. We went through the difficult process of bidding and counterbidding and eventually got the house for very little money.

The only problem was that we couldn't afford to pay both rent in Tacoma and the mortgage on the new house, but Tara had to complete the winter quarter at Pierce. Finally, we decided that Tara would commute and stay in Tacoma part-time to finish up her teaching responsibilities. This was a difficult time: a new baby, a new job, a new house, commuting, and on top of it all the fact that the whirlwind change had left me dealing with the fallout of anxiety.

However, now that I was armed with my diagnosis, I immediately grasped the advantages of expanding my coping mechanisms and came up with a detailed plan to become more well adjusted and grow into a better partner, friend, and parent. As always, I took my moral obligation to humanity very seriously, but now I was finding more efficient tools to reach out. It helped enormously just to be able to tell people I had a form of autism and to explain why I did certain things, so they wouldn't take my abruptness personally. I started to recognize my limits, too. I no longer felt deeply conflicted about needing to get away from people; I knew it was necessary for my mental health. The onus was no longer upon me to press past painful limits just for the sake of doing so. In making these discoveries and relaxing into them, I actually became better able to sustain human interaction.

CHAPTER 10

Sharing the Songs

I had always seemed to get along well with other people working with apes. I think they saw me for the paradoxically simple and complex organism that I was because they had so much practice with and empathy for primates like me. I always felt relatively safe in even larger groups of these people. My first professional home was ChimpanZoo, the branch of the Jane Goodall Institute dedicated to enriching the lives of captive apes. ChimpanZoo is based in Tucson, Arizona, but the conferences move all over the country. I attended virtually all of them. Over the years I have had an opportunity to talk or work with the majority of the most respected researchers working with captive apes.

Jane Goodall herself was an important mentor to me, as she has been for scores of other people.

She is one of those rare individuals who makes one feel that they are more than just the physical presence. Jane has a foot in two worlds: the present corporeal one, and the one that is larger than most people can see. She is a doorway. She understands this, and unlike me, she seems to draw strength from being with many people. She never stays long in one place, choosing to tirelessly travel the world trying to educate people before it is too late for the apes to survive. She will regularly give public lectures and then sit for long hours to sign books and talk with people, who literally line up around the block to see and talk to her. She won't quit until the very last person is finished.

At a ChimpanZoo Conference Jane gives a public lecture. It has become a tradition for some of us to share dinner and wine with her, telling stories late into the night. At the last conference Jane, with her serene but animated face, told us one of the amazing tales that comprise the tapestry of her well-lived life, and I thought I was in the presence of a living saint: not the kind that never makes mistakes or never falters, but the kind that gets dirty from digging down into all our hidden places and, while holding our soil underneath us, uplifts our souls to something we suspected was there but could never reach alone.

When Jane agreed to write the foreword for my first book on gorillas, it was like participating in a sacrament. Perhaps because of my autism, I don't understand the phenomenon of being impressed with people simply because they are famous. My reverence for Jane has been a thing that grew in her presence. I have had dinner with her on several occasions, led roundtable discussions of which she was a member, participated in small, intimate workshops with her, where hard scientists have broken down in tears, and watched her enthrall packed auditoriums, making each person there feel singular. She warrants my highest compliment—she is unique.

Jane is not the only researcher who has shaped my philosophy, however. I have had the honor of talking with and learning from Roger and Debbi Fouts, Virginia Landau, Sally Boysen, Lyn Miles, Penny Patterson, Duane Rumbaugh, and Sue Savage-Rumbaugh. All of them, and the apes they work with, have had a great influence on my thoughts about ape cultures and the problems we now face in regard to their preservation.

Sue Savage-Rumbaugh introduced me to Kanzi, a bonobo man whose attempt to communicate with me moved me to tears. When I first met him, he asked me to play chase, and so we ran up and down along the fence, back and forth, him with a big bonobo smile and me slipping into my natural gorilla ways. Suddenly he stopped, grabbed the lexigram board containing the symbols he uses to communicate with, made a series of gestures, and then pointed to the lexigram board. I had to explain apologetically that I didn't understand what he was saying. Sue and I discovered that he had realized I was a "gorilla," had remembered seeing videos of Koko the signing gorilla, formulated the hypothesis that gorillas use signs to communicate, and then employed *accurate* ASL signs to ask, *"You . . . gorilla . . . question?"* pointing to the lexigram for "gorilla" for emphasis. No one, not even Sue, knew he had retained signs from watching the Koko video. He had gone through all of these cognitive and emotional steps to try to bridge the communication gap between us.

I grew very fond of Kanzi, and he continued to call me "'rilla" whenever I visited. Many people don't realize that these bonobos actually speak English. It is like listening to someone with a very thick accent: once you understand the consistent patterns in their speech, you can know what they are saying with absolute certainty. Kanzi would say, *"Go! Open group room, gorilla!"* when he saw me. The first time I heard the sentence, I was delighted and surprised. It was like an epiphany—*he was speaking with me!* I had long known that apes communicate, but having a conversation with an ape, in English, made me hope beyond hope that one day soon an ape will walk up to the bench of the Supreme Court and say, "I want to be free."

Some of the things apes tell us remind us of the immediacy of their pain. Michael, Koko's companion, was able to describe through sign language the terrifying memories of his capture as a baby. Michael, tragically, died recently, but his memories remain with us.

Chantek, an orang utan who learned sign language from researcher Lyn Miles, is a gentle and giving person. He loves to make beaded necklaces and mobiles with objects that he finds intriguing. He often makes his jewelry and *assemblage* art to give to people who visit him in his current home at Zoo Atlanta. When I went to meet him, Lyn by my side, I was touched by his simple presence and generous nature. I had brought a strawberry to share with him, having wrapped it in a napkin because it was very juicy. Chantek saw the juice on the napkin and rushed over to look worriedly at my finger. *"Hurt! Hurt!"* he signed. He wouldn't relax until I showed him both sides of my hands, free of injury. I had learned some sign language, so I could talk to him a little and not appear rude.

"You always make necklaces for other people," I said. *"So this time I'm going to make one for you."* I strung some beads and tied the long length of twine. Lyn handed it to him, and he put it on his head, took it off, and gave it back to me to keep. I was deeply moved by this orang utan man, who had been through such hardships and neglect, yet had been concerned for the pain he thought I was in when I arrived. And now he wanted to give me something to take with me.

Though the gorillas are my First People, the suffering of all apes in captivity and their impending demise in the wild are intrinsically linked with the ways humans respect—or do not respect—the other lives around them. This lack of respect extends to all the apes, other animals, even plants and the elements. The conviction that we need to respect other living things runs throughout the ape advocacy movement, though most people feel that they can focus on only one species at a time, perhaps choosing the one that speaks to them, as the gorillas have to me.

Though one might optimistically believe that ape advocates would, necessarily, see eye to eye, those of us who have spent years promoting the welfare of apes, whether captive or wild, know that it is not so easy. People who feel passionately about the animal nations find that they are often strongly bound up in one particular means of solving the complex problems we face. Since the ape advocates I meet all have noble intentions, I have been inclined to believe that all of them are essentially right and have invaluable parts to play toward the peaceful resolution of these violent times. I try very hard to synthesize the strategies I come upon; when I can't, I remind myself that the dynamic tension resulting from irreconcilable approaches is the very place where new creations emerge—not necessarily through human effort, but perhaps from something more creative.

Learned people have written many volumes on the subjects we continue to discuss: Are apes in fact persons? What are human beings' obligations to them? Should we be able to use them in ways that are abusive? Do they have a right to freedom? If they do, what does that freedom look like?

In order to approach these questions intelligently, in my opinion, some historical background is important. There is a growing movement to equate the current approach to primate (and other animal) management with the human slavery practiced by the United States a century and a half ago. I agree that this is a valid comparison. The parallels are extremely compelling. Many times I have been reading a book on the justifications given for human slavery in the antebellum South or one on our current justifications for having apes in captivity, and I have had to flip to the book jacket to remember which I was reading about. The common justifications—that the beings in question are a different species, that human people benefit financially (or even medically) from their enslavement, that they are incapable of self-management, that they are far inferior on an intellectual level, that they lack souls or the capacity for spiritual understanding and enlightenment, that they are inherently dangerous and shouldn't be allowed to live freely among us, that they are clearly physically different, repugnant, and debased—currently underpin as heated a political climate and as strident a set of opinions as they did 150 years ago.

Those of us who believe that apes are unjustly seen as too different from "us" to warrant ethical consideration—we who are aware that apes are fully capable of moral agency, love, self-awareness, rational thought, and spiritual awareness—have a second historical reality to deal with: those who set out to dismantle oppressive systems that are taken for granted culturally must understand that the burden is upon them to offer a clear plan of action in regard to setting up the freedom proposed for the enslaved being of the latest round.

Permitting history to guide us in our decision making, we can generally go two ways when we consider rebutting the justifications for the enslavement of a sentient group of beings. One is to move to protect the beings in question by arguing convincingly that they are sufficiently like "us" to deserve that protection. Also let us remember that membership in that group is always changing. The second option is to allow the current cultural thinking to go unchallenged but to advance the notion that even if the beings in question aren't like us, they still suffer when abused and warrant protection under the law. In my opinion, a version of the second approach has already been adopted to varying degrees by the dominant human culture and is therefore easier to expand upon for that reason. For instance, we have organizations such as the American Association of Zoos and Aquariums and the Institutional Animal Care and Usage Committee that oversee facilities that keep animals captive. These organizations try to ensure, with varying degrees of success, that the "limited capacities for suffering" that many humans assume animals have are not being grossly violated. There is some historical evidence that this slow awakening to the feelings of other species is a necessary road human people must travel to institute permanent change. As a species, we do not work much at instituting freedoms for other beings.

The components of prejudices—such as the way we are trained to see bodies of other species as different and inferior, and our belief that they are not as smart or as evolved as we are—are changed at different rates as human people become used to the idea that their assumptions could be wrong. The by-products of these prejudices—using the other species for profit or believing that they need human people to keep from damaging themselves or others—could take generations to overcome. As we know, apes and other animals that are endangered or suffering don't have generations.

This brings us to the approach to social change for apes: the argument for "personhood" based on their similarity to ourselves. As I hope I have shown in the story of my personal transition, apes fulfill all of the criteria that currently define personhood: self-awareness; comprehension of past, present, and future; the ability to understand complex rules and their consequences on emotional levels; the ability to choose to risk those consequences, a capacity for empathy, and the ability to think abstractly. If apes warrant the bestowing of even some rights of personhood, we must address the difficult questions that face us, as we are the only species capable of granting those rights.

For example, what is the optimum habitat for captive apes? Most captive situations are not ones we would support as adequate for persons: ape people have limited say in how and where they live and are at the mercy of human caprice. In narratives written during the abolition period in the United States (from slaves' point of view), several key concepts about freedom are repeated: the importance of the control over one's physical body in regard to labor, punishment, and other harms; freedom of movement; and influence in the sphere of government—which in the case of apes could someday extend to their having a say over how their ancestral lands are used or which kinds of freedoms they deserve. It will take years to reach a social atmosphere wherein apes can enjoy all these freedoms (as human oppression of other humans has shown). But I believe we need to describe these freedoms and their importance.

Another difficult question is whether captive ape men and women should be allowed to have children. There are valid arguments on both sides of this issue. For human people who are trying to get apes out of labs and place them in sanctuaries, the ethical answer is clear: no child should be born into a lab, and there are not enough sanctuaries to house the adults currently alive, let alone larger, more natural places where they can have some freedom to raise families in a fashion that approximates their natural cultures.

This is a particular problem for chimpanzee people, as they are not considered endangered in the same way as bonobos, orang utans, and gorillas by the government, and many are in terrible forced breeding situations to keep labs stocked. A perhaps more noble argument for ape childbearing is that apes are not going to make it in the wild, so we must therefore allow them to have children in captivity. This is the position of most zoos. Indeed, if we see them as people, we can't deny them the most basic, beautiful, and life-affirming of experiences: having children. As with the other challenges we face in advancing ape freedom, there are examples that support this position. In some zoos ape families appear to be happy in that they enjoy a longstanding and intact family and the men and women love each other and naturally express their intimacy by making love and sharing children. In other contexts, ape people are constantly torn from their families so that they can "enrich the gene pool" by being forced to breed elsewhere. This practice is obviously not popular with many influential ape advocates. But when I talk to the most extreme opponents of childbearing, the main reason for their antagonism almost always comes down to their belief that the apes in question should have the physical room to replicate their cultures.

An even more basic and immediate issue is the use of apes in biomedical research. Proponents claim that the pain and suffering of apes benefits humanity to a degree that justifies the apes' loss of freedom, their isolation from other living things, their endless agony, and often their eventual death, but these arguments have been made before and are always the same: the doctor in the 1800s who, in the name of science, cut a black man open to see how deep his brown went; the doctors who let African American men go untreated for syphilis in order to see how the disease progressed unchecked. It progressed horribly. Those who experimented on the Jews of sixty years ago—injecting them with acid, decompressing them until their brains exploded, breaking the children's bones—believed that their subjects were not as important as those who would benefit from the research. I have heard arguments that things are different now: we know for *sure* that those who are experimented on aren't as important as "us." Our methods are better scientifically. The results can be applied more efficiently to those who need them. But these are the same things that have always been said when it was economically beneficial, culturally sanctioned, and morally expedient to harm others for the good of those in power.

And then there is the smiling ape used to sell products. Most of the public is unaware that the "smiles" on the faces of apes in entertainment and advertising are actually fear or stress grimaces. How the trainers teach the apes to respond this way is up to them. Trainers take infants away from their parents—who are often badly treated "breeding stock" living in warehouses—when they are tiny and still need their mother's love and social guidance. These ape children are forced to perform, are punished and sometimes beaten as they get older and explore their dominance (a number of trainers use electrical shock belts concealed under apes' clothing to continue to control them and have even beaten them to death). The orang utan used in the Clint Eastwood movie *Any Which Way But Loose* was severely beaten with an ax handle prior to filming and died within a month of the movie's release.

Trainers eventually "retire" apes at around seven years old, selling them into research or warehousing them to provide a new generation of show apes.

There are no quick or easy answers.

Some human people, like the repatriation advocates within the abolition movement of the nineteenth century, will settle for no less than repatriation of all captive ape people to Africa. My own position, and that of some of my researcher friends, is that we have created bicultural beings, and it would be difficult, if not impossible, to take them all back to Africa and pretend we didn't take them out in the first place. Beyond the philosophical problems that repatriation might prove to have, there is the pragmatic reality that no habitat exists to which we could repatriate them even if we wanted to. The habitat that does exist is overrun by the bushmeat trade, where apes are slaughtered and sold for meat—one can buy a smoked gorilla for twenty dollars in the Congo. These problems are pushing the apes already there into extinction. I agree that ape families should have the right environment to raise their children, but it is up to us to provide that in a context that does not spell sure doom for them.

In the final analysis I believe it is our responsibility to supply all the apes we now hold captive with all that they need to retain their cultures and to evolve. I urge people who want to do something to aid in promoting the well-being of apes, and as a result all other living things, to find an organization that reflects their particular views and support the inspiring work being done. My own personal choices include the Great Ape Project, the Jane Goodall Institute (and ChimpanZoo specifically), and the Primate Freedom Project.

One of the most visionary endeavors that I support is being undertaken by Penny Patterson and Koko, in partnership with Lyn Miles and Chantek, and will eventually include many other captive apes. In their project, aided by the support of the Gorilla Foundation and ApeNet, apes of all tribes who are currently involved in language research across the country will be able to speak to one another through live visual feeds. Additionally, the Gorilla Foundation is working to establish an immense sanctuary on Maui, where the apes can come and live; they would talk to one another and human people in real time, they would have the physical space and resources to perpetuate and expand their cultures, and they would enjoy the kind of freedom that will allow them to reach their full potential, perhaps even to teach humans about their own. The work is currently underway to accomplish this amazing step in interspecies communication—which will bring humans, apes, and perhaps other animals together through language. I consider it the most significant effort to date in the goal to connect sentient beings, but funding is still needed.

I hope that readers of my story will understand the great gift that the people of the ape nations are: not for what they can do for us, not for what they have done for me, but for who they are in and of themselves. Whatever group the reader chooses to support, I urge them to become involved.

Epilogue

My life and career have grown into something I could never have dreamed of so many years ago. I know how to show people my best side and communicate my skills while knowing when to stop. To become an adjunct professor in the anthropology department at my university, I used a formula: a third of the time I talked about my skills, another third I talked about the work of my colleagues and how our interests merged, and the final third I talked about current events and my (softened) opinions about them. I remembered to ask questions about other people's interests.

Unlike other people, I often consciously think about these divisions of topic in a conversation. Many times I also count how many seconds to look into someone's eyes and how many to look away. I try to take in all that is being said and remember it, all the while trying to read others' emotions and respond in a way that makes them feel good. This skill has become easier with time. I have used it to show my interest in the work of people I admire greatly, and I now work with some of the best-known people in my field.

Even though I have made a lot of progress in dealing with my fears and the panic I feel over situations, there are still those that manage to incapacitate me. One is trying to figure out how a vending machine works. Usually, I am already tense from being in a public place. If I feel as though people are watching me, I start to get the tunnel vision that so many autistic people describe. For example, I went to a local copy store today, knowing that they had a FedEx station there, as I needed to mail in some page proofs to a publisher. I took my son with me. When we went in the door, I noticed it was crowded. Oh, boy. I looked around for the FedEx station, though I had seen it many times before.

"What are we looking for, Mom?" my son asked.

"I'm looking for a box with a bunch of envelopes in it. I think there is an eagle on the side," I said, my mouth dry.

My son took my hand and said, "Is this it?" as he led me to the station. I thanked him. Then we stood for long minutes as I tried to work out what I was supposed to do. With my vision the size of a dime, I moved my head around to find the directions. There were directions for filling out the waybill, selecting an envelope, et cetera, but none for how to pay. I looked all over the box, feeling a rising sense of animal panic. Finally I just stood dumbly as the box virtually disappeared altogether. My son pulled at my leg. "What are we *doing*," he whined. "We're going to find a place that does UPS," I said, and picked him up, beating a hasty retreat to the car, where I took a minute to get my bearings.

What is worse is when I find a vending machine that

takes my money and I have to make the dreaded decision about whether to go find someone to get it out for me. I almost always walk away, feeling it is worth fifty cents to avoid dealing with a stranger.

There are other things that scare me that are actually enjoyable to most other people. Clowns, for instance. The average human face is enough of a challenge; most of them seem overdone already to a lot of autistic people. It is hard to express the horror I feel when I am out at a parade or carnival (already a sensory nightmare) and I see a clown coming. The garish colors of an exaggerated smile, the electric daggers that are rainbow wigs, the oversized hands and feet: all of these make me want to run at top speed for the nearest exit. If I can't get away, I sometimes feel like I want to attack the clown. This is a social no-no.

What I can't for the life of me figure out is what a person is supposed to do when approached by a clown. What is so enjoyable about this experience? As far as I can tell, it is unprecedented in nature to peacefully allow something that bright and colorful to come at you. Some things I'll never understand.

Another thing that can really throw my day is seemingly very small, but its impact is great. It is another thing that most people find pleasant: a friend honking hello to you as you walk down the street. This wrecks me every time. When I walk, I think. I get lost in the slow passing of the scenery, almost hypnotized, to a point where I feel very relaxed and happy as my deepest thoughts emerge and I follow them wherever they lead. Often, I think about what I am going to write next; I usually compose something entirely in my head, even editing it there, then all I have to do is transcribe it onto paper (I rarely make revisions once I have written). So I'll be happily stewing in my thoughts, arranging and rearranging my thoughts—my world—when *honk!* The peace is shattered by a careening streak of metal and glass, hitting me broadside and blasting me out of myself. There is often the bonus of the person who knows me leaning out of the car screaming hello, one hand on the wheel of their speeding carriage, the other, with varying amounts of their body attached, leaning out of the window to wave at me wildly. I love my friends, and I know they do this to be nice and that they are excited to see me. I have learned to fake immediate recognition and wave back enthusiastically in the general direction of the commotion. Often I find out later who it was, but in the meantime, I remain shaken up, my heart racing, my head foggy. I keep walking, trying to remember where I was going. Throughout the day I will play the series of events—the noise, the colors, the shouts—as if they were a video on an automatic loop.

I find comfort in collecting things: toy monsters and old horror movies, fossils, model brains, and antiques that remind me of my childhood. I continue to have chronic insomnia and panic attacks. I often must move my legs or feet constantly and vigorously to keep myself "intact."

I like to wear the same clothes over and over and hate to wash my favorite pair of pants. Like my mother, I often wash them and dry them at night so that I can retrieve them from the dryer immediately in the morning. I can actually feel my shoulders lose their tension as I slide my pants on. When I speak publicly or at conferences, I must always wear the clothes I will wear to bed the night before and sleep in them to condition them properly. Also when I am anxious, I wear my clothes to sleep in at night.

I never experience physical sensations of hunger and thirst, and as a result I am often dehydrated, which leads to headaches and dizziness. I sometimes clear my throat obsessively in social situations. I still can't recognize faces and have to meet people in front of restaurants or other gathering places so that they can find me before I have to try to find them. I leave an hour early to go to new places, knowing that I may get lost. After social gatherings I often must debrief with my partner or friends in order to be sure I understood the implications of things that were said or events that occurred.

I have learned a lot from sharing a life with Tara. She has taught me to trust through change and to believe that flexibility is important to creativity. Though I still dislike chaos, what I experience as chaos has changed over time. Instead of panicking if the house isn't completely organized, I have come to feel that the house is a living, breathing entity, an organic process that enfolds our family and reflects it like a net of jewels. The dirty pots on the stove become evidence that contented people with the gift of bounty have shared something sustaining and tangible. The scattered garden tools in the front yard are artifacts of a hot, passing day full of new growing things and a promise of flowers and strawberries. I see Teryk's toys strewn on the floor, and I think about what it would be like if they weren't there at all. Where I used to need control over the placement and care of everything in the house, I have relaxed about giving it over to Tara and our son.

I have learned that when people feel an emotion, it will pass. I used to become terror-stricken when disagreements, anger, or sadness happened because I simply couldn't imagine that something different would follow, whether it was with myself or others. I would get stuck in the emotion as if it were some kind of molasses. My mind would circle back in on itself again and again. It was always very frightening to think I would feel angry forever, or sad forever, or confused forever. Tara, by forcing me to engage in the *processes* of different emotions, showed me that I could calm down and think about how to direct what happens next. I also gained some insight into the ways I took things said at face value, as completely literal. I used to get defensive and hostile before I realized playful teasing isn't meant as criticism. I still have trouble with this.

When I realized how difficult I had made Tara's life, I was ashamed. My first thought was, *What would Congo think of me? I have failed as an important force in a family.* As I continue to grow in my relationships with Tara and other people I love, I have tried to be more like him. Because I need a set of absolute rules in order to function and do what I believe is morally correct, I take as my code silverback ethics and a sense of gorilla responsibility. I remember that my mood can set the tone for a whole group of people. I realize that it is my responsibility to protect and nurture those around me, my family in particular, so that they can achieve their full potential. I remember to be gentle in the reality of the force that an interesting and difficult life has given me. I affect people.

Nowhere is this as obvious as with my son. He, in his limitless openness and love, absorbs everything I do and say. This fact overwhelms me at times. He so easily walks between the real world and the magical worlds that children can touch just by holding out their hands. He reminds me of the ways my worlds as a child, though sometimes hard to manage, were also magical. I can often follow him into his worlds more easily than other people can. It is wonderful to have a guide as I enter those worlds as an adult, and my little son takes me by the hand to show me where colors dance around the room, where mythical birds rest in the trees, and where water spirits silently weave lilies for the day to find. Smiling up at me, he gives me simple answers to problems I overthink. He gives me beauty as if it were a present that he could put in a box.

When he was born, I held his slippery little body in my hands and, sobbing, wondered if he would remember being born. He says he does, and I believe him.

Other things about being a parent aren't so rewarding. For example, it is understood that I will interact with the parents of the temporary friends he meets on the playground:

Strange parent: That's some boy you've got there.

Me: Yeah, I think I'll keep him. (This always gets a laugh, so I have adopted it as a script line.)

Strange parent: How old is he?

Me: Three. How old is yours? (This last part is required, so don't try to get out of it.)

Strange parent: I sure wish I could bottle their energy!

Me: Yeah, we'd make a fortune. (Stranger chuckles, and then there is a silence I never know what to do with.)

The conversation goes on in this way until I walk off to play with my son, which is usually fairly quickly. I hate these conversations. I am just bursting to ask what the stranger thinks of the idea of folded space and superluminal communication, or the possibility that a common hominoid ancestor could have been bipedal previous to the ape-human split, or the phenomenon of resonant formative causation. Sometimes I think I'll die if I hear another parent relate the way their daughter likes to dress up as a princess.

This challenge is carried over when I have to deal with his teachers. (Luckily, I like all of them.) I walk into preschool, and I'm immediately assaulted by loud children, loud colors, and loud smells. Feeling my vision start to tunnel, I look frantically for my son's face. I am always so happy to see him. My vision opens up as I hold him and ask about his day at school. Then I know I need to talk to his teachers—not to do so would be considered rude and uncaring. Each time I'm not sure what to ask. I usually try some standard questions: What did he do today? Did he do any artwork? Did he have fun on the playground? Was he cooperative? It isn't that I don't care. I would just rather hear it from him in a quieter place where both of us can really talk. I think he is a better judge of what kind of day he had, what he did—what was important to him. The most important thing I have learned is that his needs come first, and his thoughts are dear to me. I want to be able to really listen. I think being able to listen to him has taught me how to listen to myself.

I still start a significant number of my conversational offerings with "From a primatological point of view . . ." or "Anthropologically speaking . . ." Thankfully, I have found a home in the academy. It has placed me among people who are often open to the bright and eccentric, to the new and different, and who are interested in new ways of perceiving and experiencing. My involvement in the university has allowed me to thrive in an atmosphere where thinking (at least to many people) is foundational to living. In many ways my autism and my unique history as a learner have become assets.

Though my academic career has been vastly different from that of anyone else I know, people have been intrigued by and envious of the fact that I worked alone under the direction of mentors rather than through more traditional means that many times do not work for really bright students. But though I am happy and well suited to my position as a researcher and writer—a position that allows me to focus on research and mentor students individually—I am also aware that my autism and the academic path it required have kept me from gaining a tenure-track professorship and attaining security.

Autistic people who do not have an understanding of their condition are often able to find niches that allow them to use their talents and retain a certain amount of control, but these jobs are often low-paying, void of proper recognition, and basically not secure, having little room for advancement. This can occur for several reasons, including poor interview skills due to anxiety (I had one interview, and it didn't go very well), the inability to select and prioritize pertinent information on a written application, and an aversion to the social situations that jobs present. In my case, I think my alternative education was the first hurdle—only now are job candidates with interdisciplinary degrees being accepted and even sought out. Though I can choose and prioritize the material I teach without any problem at all, it is difficult for me to know for certain what a search committee is looking for. There are many unwritten rules in writing an application and the requisite cover letter. An autistic person might think, for example, *Well, it's clear from my résumé that I'm qualified and how I got that way. They probably will want to know more about who I am, since we will have to work together for many years if I'm hired.* Based on this line of thinking, the autistic applicant will complete a very personal (and inappropriate) cover letter. Because of it they won't make it past the first screening.

A mistake I used to make when I was actively seeking university jobs was to send potential employers and colleagues work in its early stages, sometimes even preliminary notes, in order to share my projects with them. My reasoning was that if I respected and admired someone, I would want her or his input at the very beginning of a project so that we could discuss even our most practical assumptions *as a process together.* It was my way of trying to really connect deeply with people I liked. I thought such sharing would give a hiring committee insight into my style as a scholar and my thinking processes. Now I realize that people see this as unprofessional behavior. This state of affairs continues to make me sad, as it seems like a very limiting approach to creating and knowing.

Despite these continued difficulties, I feel like a bridge between my ancestors and the next generation, between autistic people and the "normal" world, between apes and humans, between what is possible to change and what is not. Maybe all people who come close to reaching their full potential feel this way. It's like knowing a song so well, you don't have to think it to sing it, and all the while other people hear it and think it's beautiful. The songs of the Gorilla Nation are like that, and I sing them every day. I will never have to sing them alone.

It is my hope that through this book not only students of anthropology and autism but everyone else as well will see gorillas as teachers, too. It is my hope that we will all become students of the gorillas' gentle care, fierce protectiveness, love, and acceptance. Perhaps if human people learn these same things, if we truly learn these lessons together, then "a culture of one" will mean a culture of all.

As a writer, as an autistic person, and as a researcher of primal identity and culture, I find that the binding reality of elemental forces and the embodiment of our archaic origins have strongly influenced my views on the nature of the autistic condition. Though their research is rarely if ever cited anymore, I agree with Elisabeth and Niko Tinbergen, the authors of *"Autistic" Children: New Hope for a Cure,* who believe that modern life, with its unnatural living conditions, chemicals, broken-down social systems, and chronic stress, overstimulates and assaults the human animal, causing some to manifest the biological and psychological matrix we call autism.* They charge that as animals, we engage in all kinds of behaviors that, in autistic people, become tight feedback loops almost impossible to get out of. *There are conflicted feelings of wanting to*

*But there is at least one possible case of autism in a wild chimpanzee. Many people remember Flint and Flo of Jane Goodall's landmark study. Flint engaged in perseverative behavior, had poor impulse control, was unusually aggressive, lacked social skills, and was hypersensitive to change. He continued to nurse on his mother and ride on her back for many years past the age when most young chimpanzees are weaned and fairly autonomous. When she died, he was unable to function without his mother as an intermediary force and died of grief and stress three weeks later.

both approach and avoid people and situations, inhibited intentions and movements, rituals incorporating displacement and even self-injury, and, of course, oversensitivities to stimulation. Zoo animals without adequate context do all of these things. I have often felt like them: an anxious animal in a zoo.

Autism is a way of sensing the world—the whole world—of creating and knowing. It is my hope that as more autistic people find places to learn about themselves and grow, as they tell their stories for themselves and all people, they will find ways to share their special talents with the world. I hope they will be perceived as being as whole as the worlds they sense.

Like all human and other persons, we are not only part of things but whole already. As whole cultures within one, we have much to sing about.

Additional Reading

Attwood, Tony. *Asperger's Syndrome: A Guide for Parents and Professionals.* London: Jessica Kingsley Publishers, 1998.

Frith, Uta. "Asperger and His Syndrome." *Autism and Asperger Syndrome,* edited by Uta Frith. Cambridge: Cambridge University Press, 1991.

Grandin, Temple. *Thinking in Pictures: and Other Reports from My Life with Autism.* New York: Doubleday, 1995.

Grandin, Temple, and Margaret Scariano. *Emergence: Labeled Autistic.* Tunbridge Wells: Costello, 1986.

Miedzianik, David. *My Autobiography.* Nottingham: Child Development Research Unit, University of Nottingham, 1986.

Tantam, Digby. *Eccentricity and Autism.* Unpublished Ph.D. diss., University of London.

Tinbergen, Elisabeth and Niko. *"Autistic" Children: New Hope for a Cure.* London: George Allen & Unwin, 1983.

About the Author

Dr. Dawn Prince-Hughes received her M.A. and Ph.D. in interdisciplinary anthropology from the Universität Herisau in Switzerland and is an instructor in the department of anthropology at Western Washington University. She is the author of *Gorillas Among Us: A Primate Ethnographer's Book of Days,* the editor of *Exceptional Admissions: Essays by Autistic University Students,* and the executive chair of Ape-Net, a nonprofit organization.